LONDON MATHEMATICAL SOCIETY LECTURE NOTE SERIES

Managing Editor: PROFESSOR I. M. JAMES, Mathematical Institute, 24-29 St Giles, Oxford

This series publishes the records of lectures and seminars on advanced topics in mathematics held at universities throughout the world. For the most part, these are at postgraduate level either presenting new material or describing older matter in a new way. Exceptionally, topics at the undergraduate level may be published if the treatment is sufficiently original.

Prospective authors should contact the editor in the first instance.

Already published in this series

T0297315

London Mathematical Society Lecture Note Series. 33

Permutation Groups and Combinatorial Structures

N.L.BIGGS

Reader in Pure Mathematics
Royal Holloway College
University of London

A.T.WHITE

Associate Professor of Mathematics
Western Michigan University

CAMBRIDGE UNIVERSITY PRESS
CAMBRIDGE
LONDON · NEW YORK · MELBOURNE

CAMBRIDGE UNIVERSITY PRESS
Cambridge, New York, Melbourne, Madrid, Cape Town, Singapore, São Paulo

Cambridge University Press
The Edinburgh Building, Cambridge CB2 8RU, UK

Published in the United States of America by Cambridge University Press, New York

www.cambridge.org
Information on this title: www.cambridge.org/9780521222877

First published 1979
Re-issued in this digitally printed version 2008

A catalogue record for this publication is available from the British Library

Library of Congress Cataloguing in Publication data
Biggs, Norman.
    Permutation groups and combinatorial structures.

    (London Mathematical Society lecture note series; 33)   Edition
    published in 1971 under title: Finite groups of automorphisms.
    Includes bibliographical references and index.
        1. Permutation groups.   2. Combinatorial analysis.   I. White,
Arthur T., joint author.   II. Title.   III. Series: London Mathematical
Society. Lecture note series; 33.
QA171. B564   1979   512'.2   78-21485

ISBN 978-0-521-22287-7 paperback

Contents

Introduction

This book replaces an earlier one, entitled 'Finite Groups of Automorphisms' (No. 6 in the L.M.S. Lecture Note Series). When it became clear that the original book was still in demand, it was decided that a complete revision was preferable to a reprint. In this way we hoped to incorporate the fruits of experience, gleaned from readers' comments, and at the same time, to keep the book up-to-date by introducing some new material.

The entire book has been rewritten. In some sections of Chapters 1, 2 and 3 the earlier version has been followed quite closely, while in others (notably 1.5, 1.6, 2.4, 2.5, 3.4, 3.5), the material is treated differently. Chapter 4 (Groups and Graphs) is presented from a new viewpoint, beginning with the graphical representation of permutation groups. The algebraic theory of the adjacency matrix has been restricted to the case of strongly regular graphs, since the generalization to graphs of larger diameter is discussed elsewhere (reference [1] of Chapter 4). The material on the Higman-Sims group has been amplified. Chapter 5 (Maps) is completely new. We hope that its inclusion will lead to renewed interest in a subject where ideas from many different areas of mathematics come together.

We have tried to make the book suitable for use as a course text at advanced undergraduate or postgraduate level. For this reason, there are thirteen 'project' sections, which should provide a good test of a student's real understanding of the text. The topics treated in the project sections are occasionally needed in subsequent chapters. (See also the note below.)

We are grateful to our colleagues who, from January to March 1978, attended a weekly study group on these topics at Royal Holloway College; their support was much appreciated. Also, we received helpful comments and suggestions from several people, including P. J. Cameron,

A. D. Gardiner, and W. M. Kantor. Dr Gardiner's help has resulted in substantial improvements to Sections 1.5, 3.7 and 4.6. Finally, our thanks are due to Professor I. M. James for his encouragement, and to the splendid staff of the Cambridge University Press.

NOTE ON THE PROJECTS

The projects are not routine exercises. They are intended to be substantial pieces of work and, typically, a student might expect to spend several days working on each one. It might also be necessary for him to seek help from books, fellow-students, or his teacher. It is probable that no student will have time to attempt all the projects, but it would be an advantage to be familiar with the major results contained in them. Some parts of the projects are starred: these are either quite hard, or very time-consuming.

1 · Permutation Groups

'... it will afford me much satisfaction if, by means of this book, I shall succeed in arousing interest among English mathematicians in a branch of pure mathematics which becomes the more fascinating the more it is studied. '

W. Burnside, in his preface to Theory of groups of finite order, 1897.

1.1   Preliminary definitions

It is presumed that the reader will already be familiar with the contents of this section. He is advised to read it quickly, in order to accustom himself to the notation which will be used throughout the rest of the book.

If X is a finite set, a permutation of X is a one-to-one correspondence (bijection) $\alpha : X \rightarrow X$. Two such permutations α, β can be composed to give the permutation $\alpha\beta : X \rightarrow X$, which we shall define by the rule $\alpha\beta(x) = \alpha(\beta(x))$. That is, we shall write functions on the left, and compose in the order compatible with this convention. Under the operation of composition the set of all permutations of X forms a group $Sym(X)$, the symmetric group on X. If X is the set $\{1, 2, \ldots, n\}$, we write S_n for $Sym(X)$, and we have $|S_n| = n!$ where we use $|\ |$ to denote cardinality.

If G is a subgroup of $Sym(X)$, then we shall say that the pair (G, X) is a permutation group of degree $|X|$, and that G acts on X. More generally, we shall meet the situation where there is a homomorphism $g \mapsto \hat{g}$ of a group G into $Sym(X)$: this will be called a permutation representation of G. When the homomorphism $g \rightarrow \hat{g}$ is a monomorphism, we say that the representation is faithful; in this case it is convenient to identify G with its image in $Sym(X)$, so that we recover the case of a permutation group (G, X).

Any permutation α in $Sym(X)$ may be decomposed, in an essentially unique way, into disjoint cycles:

$$\alpha = (a_1 a_2 \dots a_k)(b_1 b_2 \dots b_l) \dots .$$

This notation means that $\alpha(a_1) = a_2$, $\alpha(a_2) = a_3$, $\dots$ $\alpha(a_k) = a_1$, and so on. A <u>transposition</u> is a cycle (ab); since $(x_1 x_2 \dots x_r)$ is equal to the composite $(x_1 x_r)(x_1 x_{r-1}) \dots (x_1 x_2)$, any permutation can be expressed as the composite of (not necessarily disjoint) transpositions. This expression is not unique, but if

$$\alpha = \tau_1 \tau_2 \dots \tau_k = \sigma_1 \sigma_2 \dots \sigma_l,$$

where the τs and σs are transpositions, then $k \equiv l$ (mod 2). When $k \equiv 0$ (mod 2) we say that α is <u>even</u> and write $\text{sgn}(\alpha) = 1$, and when $k \equiv 1$ (mod 2) we say that α is <u>odd</u> and write $\text{sgn}(\alpha) = -1$. The sgn function is a homomorphism of $\text{Sym}(X)$ into the multiplicative group $\{1, -1\}$ and its kernel, the set of even permutations, is a normal subgroup of $\text{Sym}(X)$, known as the <u>alternating group</u> $\text{Alt}(X)$. We write A_n for the alternating subgroup of S_n. It is easy to see that

$$|\text{Sym}(X) : \text{Alt}(X)| = 2, \quad \text{and} \quad |A_n| = \tfrac{1}{2}n! .$$

If α and π are in $\text{Sym}(X)$, and α has the cycle decomposition given above, then

$$\pi \alpha \pi^{-1} = (\pi a_1 \pi a_2 \dots \pi a_k)(\pi b_1 \pi b_2 \dots \pi b_l) \dots .$$

We say that $\pi \alpha \pi^{-1}$ is the <u>conjugate</u> of α by π: two elements of $\text{Sym}(X)$ are conjugate in $\text{Sym}(X)$ if and only if they have the same cycle shape. Thus the number of conjugacy classes in S_n is just the number of partitions of n. For example, in S_5 we have

Partition	Class representative	Number in class
1,1,1,1,1	Identity	1
1,1,1,2	(12)	10
1,1,3	(123)	20
1,4	(1234)	30
5	(12345)	24
1,2,2	(12)(34)	15
2,3	(12)(345)	20
		$\overline{}$
		$120 = 5!$.

Here, as is customary, we have suppressed cycles of length one.

1.2 Counting principles

A basic technique of combinatorial mathematics is to count the same set in two different ways and equate the answers. Precisely, let U and V be finite sets, S a subset of $U \times V$, and define

$$S(a, \cdot) = \{v \in V \,|\, (a, v) \in S\},$$
$$S(\cdot, b) = \{u \in U \,|\, (u, b) \in S\}.$$

Then $S(a, \cdot)$ is in one-to-one correspondence with the subset of S consisting of the pairs (u, v) with $u = a$, and these subsets, for $a \in U$, partition S. The analogous result holds for the sets $S(\cdot, b)$. Hence

1.2.1 $\quad |S| = \sum_{a \in U} |S(a, \cdot)| = \sum_{b \in V} |S(\cdot, b)|$.

If, in a particular example, we can show that $|S(a, \cdot)| = r$, and $|S(\cdot, b)| = s$, independent of a and b respectively, then it follows that

1.2.2 $\quad r|U| = s|V|$.

We apply this method to the action of the permutation group G on a finite set X. Introduce the temporary notation

$$G(x \mapsto y) = \{g \in G \,|\, g(x) = y\} .$$

3

We check that the statement "G(x ↦ y) is not empty" defines an equivalence relation on X, and denote the equivalence class of x by Gx. If $y \in Gx$, then choosing any g in $G(x \mapsto y)$ gives us a one-to-one correspondence:

$$G(x \mapsto x) \longleftrightarrow G(x \mapsto y) \quad \text{defined by} \quad h \longleftrightarrow gh.$$

Now the counting principle applies: fix x in X and let P_x denote the subset of $G \times X$ consisting of those pairs (g, y) for which $y = g(x)$. Then $P_x(\cdot, y)$ is just $G(x \mapsto y)$, and so

$$|P_x(\cdot, y)| = \begin{cases} |G(x \mapsto x)| & \text{if } y \in Gx, \\ 0 & \text{if } y \notin Gx. \end{cases}$$

Also, $|P_x(g, \cdot)| = 1$ for all g in G. Hence 1.2.1 implies that

$$1.2.3 \quad |G(x \mapsto x)| \, |Gx| = |G|.$$

This is the fundamental relation of the theory of permutation groups. The set Gx is called the <u>orbit</u> of x, and $G(x \mapsto x)$ is called the <u>stabilizer</u> of x - it is a subgroup of G and is usually written G_x. Thus 1.2.3 becomes

$$1.2.4 \quad |Gx| = |G : G_x|.$$

In order to find the number of orbits of G on X, we may use a formula involving the set $F(g)$ of <u>fixed points</u> of g; that is, the set of all x for which $g(x) = x$. The result is often called 'Burnside's Lemma', although its origin is with Frobenius.

1.2.5 Theorem. <u>Let</u> t <u>denote the number of orbits of</u> (G, X). Then

$$t|G| = \sum_{g \in G} |F(g)|.$$

Proof. Let $E = \{(g, x) \in G \times X \,|\, g(x) = x\}$. Then

$$E(g, \cdot) = F(g), \quad E(\cdot, x) = G_x.$$

Applying the counting principle 1.2.1, we get

$$\sum_{g \in G} |F(g)| = \sum_{x \in X} |G_x| .$$

Now let $x_1, x_2, \ldots, x_t$ be representatives of the t orbits. If x belongs to the orbit Gx_i, and g is any member of $G(x \mapsto x_i)$, then the stabilizer G_x is simply $g^{-1} G_{x_i} g$, so that $|G_x| = |G_{x_i}|$. Thus we may collect the like terms on the right-hand-side above, yielding

$$\sum_{g \in G} |F(g)| = \sum_{i=1}^{t} \sum_{x \in Gx_i} |G_x|$$

$$= \sum_{i=1}^{t} |Gx_i| |G_{x_i}|$$

$$= \sum_{i=1}^{t} |G| \qquad \text{(by 1.2.4)}$$

$$= t|G|. \quad /\!/$$

For example, let $X = \{1, 2, 3, 4\}$ be the set of corners of a square, given in clockwise order, and let $G = D_8$ (dihedral group of order 8) be the permutations of X which can be realised by rotations or reflections in a plane. The elements of G, and their numbers of fixed points, are as follows:

g	:	Identity	(13)	(24)	(13)(24)	(1234)	(1432)	(12)(34)	(14)(23)
$|F(g)|$	:	4	2	2	0	0	0	0	0

Hence $t = (1/8)(4 + 2 + 2) = 1$, and there is just one orbit.

1.3 Transitivity

1.3.1 Definition. The permutation group (G, X) is <u>transitive</u> if there is just one orbit in the action of G on X.

We may use Burnside's lemma, as in the example at the end of the previous section, to check whether a given permutation group is transitive. A more direct method is to pick one element x of X and search for elements of G taking x to every other element y in X. In the example, the existence of the permutation (1234) (and its powers) is sufficient to ensure transitivity. We remark that, in the transitive case, the basic results 1.2.4 and 1.2.5 become

1.3.2 $|X| = |G : G_x|, \quad |G| = \sum_{g \in G} |F(g)|$.

We now consider the action of G_x on X.

1.3.3 **Theorem.** <u>Suppose</u> (G, X) <u>is transitive and let</u> $r(x)$ <u>denote the number of orbits of</u> G_x <u>on</u> X. <u>Then</u>

$$r(x)|G| = \sum_{g \in G} |F(g)|^2 ,$$

<u>and</u> $r(x)$ <u>is independent of</u> x.

Proof. From Burnside's lemma 1.2.5 we have

$$r(x)|G_x| = \sum_{g \in G_x} |F(g)| .$$

The right-hand side of this equation is just the cardinality of the set of pairs $\{(g, w) | g \in G_x, g(w) = w\}$. For any $y \in X$ there is a one-to-one correspondence between this set and the set $\{(k, z) | k \in G_y, k(z) = z\}$, defined by $(g, w) \mapsto (hgh^{-1}, h(w))$, where h is any element of G such that $h(x) = y$. Thus the right-hand side of the equation is independent of x, and since $|G| = |X| |G_x|$ we have

$$r(x)|G| = r(x)|X| |G_x| = \sum_{x \in X} \sum_{g \in G_x} |F(g)| .$$

Inverting the order of the double sum, we get

$$r(x)|G| = \sum_{g \in G} \sum_{x \in F(g)} |F(g)| = \sum_{g \in G} |F(g)|^2 .$$

Since the right-hand side is independent of x, so is $r(x)$. //

1.3.4 **Definition.** The <u>rank</u> r of the transitive group (G, X) is the number of orbits of G_x on X.

In the action of D_8 on the corners of a square we have

6

$r = (1/8)(4^2 + 2^2 + 2^2) = 3$. The three orbits of the stabilizer of 1 are $\{1\}$, $\{2, 4\}$, and $\{3\}$.

The case $r = 2$ is especially interesting - here, G_x is itself transitive on $X - \{x\}$, and it follows that there is an element of G taking any distinct pair of elements of X to any other pair. In this vein, we make the following definition.

1.3.5 **Definition.** The permutation group (G, X) is k-<u>transitive</u> $(k \geq 1)$ if, given any two ordered k-tuples $(x_1, \ldots, x_k)$, $(y_1, \ldots, y_k)$ of distinct elements of X, there is some g in G such that

$$g(x_i) = y_i \qquad (1 \leq i \leq k).$$

Clearly, a k-transitive group is also l-transitive, for $1 \leq l \leq k$. Usually, when we say that G is k-transitive on X we imply that k is the largest integer for which this is so. A k-transitive group with $k \geq 2$ is a transitive group of rank two, and we sometimes use the general term <u>multiply transitive</u> in this case; the fixed point formulae 1.2.5 and 1.3.3 become

$$|G| = \sum |F(g)|, \quad 2|G| = \sum |F(g)|^2.$$

The determination and construction of multiply transitive groups is facilitated by the following lemma.

1.3.6 **Lemma.** <u>Suppose that</u> G <u>is known to be transitive on</u> X. <u>Then</u> (G, X) <u>is</u> k-<u>transitive if and only if</u> $(G_x, X - \{x\})$ <u>is</u> (k-1)-<u>transitive</u>.

Proof. Let us suppose that G_x is (k-1)-transitive on $X - \{x\}$. Given any two ordered k-tuples $(x_1, \ldots, x_k)$ and $(y_1, \ldots, y_k)$ of distinct elements of X, we may select g_1, g_2 in G and h in G_x with the properties

$$g_1(x_1) = x, \qquad g_2(y_1) = x,$$
$$h[g_1(x_i)] = g_2(y_i) \qquad (2 \leq i \leq k).$$

Then $g_2^{-1} h g_1$ is an element of G transforming the ordered k-tuples as required. The converse is straightforward. //

Thus, to determine if a given group is multiply transitive, we must examine the successive stabilizers G_x, $G_{xy} = (G_x)_y$, and so on. If G is k-transitive on X, and $|X| = n$, then repeated application of the first part of 1.3.2 yields the useful result

1.3.7 $|G| = n(n - 1)(n - 2) \ldots (n - k + 1)|G_{x_1 x_2 \ldots x_k}|$,

where the group on the right-hand side is the pointwise stabilizer of x_1, x_2, ..., x_k. In particular, we note that the order of G must be divisible by $n(n - 1)(n - 2) \ldots (n - k + 1)$. If G is k-transitive, and the identity is the only permutation fixing k points, then G is said to be <u>sharply</u> k-transitive, and its order is exactly $n(n - 1) \ldots (n - k + 1)$.

The idea of sharp transitivity is especially important in the case $k = 1$; the group G is then said to be <u>regular</u> on X, and we have $|G| = |X|$. In fact, G and X are in one-to-one correspondence, defined by $g \longleftrightarrow g(x_0)$, for any fixed x_0 in X.

1.3.8 Theorem. S_n <u>is n-transitive, and</u> A_n <u>is (n-2)-transitive, in their actions on the set</u> $\{1, 2, \ldots, n\}$ $(n \geq 3)$.

Proof. The first part is obvious, since S_n contains all permutations of the n-set. In the alternating case, we may proceed by induction. When $n = 3$, A_3 contains (123) and so it is 1-transitive. The stabilizer of the symbol n in A_n is A_{n-1}, and so, by 1.3.6 the induction step is valid. It remains to be shown that A_n cannot be more than (n-2)-transitive. To see this, we remark that the only permutation of $\{1, 2, \ldots, n\}$ which takes the ordered (n-1)-tuple $(1, 2, \ldots, n-2, n-1)$ to $(1, 2, \ldots, n - 2, n)$ is the odd permutation $(n - 1 \ n)$, which is not in A_n. Thus A_n is not (n-1)-transitive. //

We conclude this section with some remarks on the coset decomposition of transitive and multiply transitive groups.

The basic argument of Section 1.2 may be expressed in the following way: if g, g' ϵ G and x ϵ X, then

8

$$g' \in gG_x \iff g'(x) = g(x) .$$

Thus the cosets of G_x in G are precisely those sets $G(x \mapsto y)$ which are not empty. When G is transitive a complete set of coset representatives is any family $\{g_y\}$ $(y \in X)$ such that $g_y(x) = y$. We then have

$$G = G_x \cup \left(\bigcup_{y \ne x} g_y G_x \right) .$$

When G is multiply transitive there is another decomposition, in terms of the <u>double cosets</u> $G_x g G_x$.

1. 3. 9 Lemma. <u>Suppose that</u> G <u>is k-transitive on</u> X $(k \ge 2)$ <u>and</u> $g \notin G_x$. <u>Then</u> $G = G_x \cup G_x g G_x$.

Proof. Let h be any element of $G - G_x$. Since G is multiply transitive, there is some g_1 in G taking $g^{-1}(x)$ to $h^{-1}(x)$ and fixing x. This implies that $hg_1 g^{-1}$ belongs to G_x, that is, h belongs to $G_x g G_x$. //

This lemma will be useful in the construction of multiply transitive groups.

1. 4 Applications to group theory

An abstract group G acts on the set 2^G of subsets of G in two especially important ways:

(i) $g(K) = gK$

(ii) $g(K) = gKg^{-1}$ $\qquad (g \in G, \ K \subseteq G) .$

Many notions and theorems of group theory can be expressed in terms of these actions. For example, if K is a subgroup of G $(K \le G)$ then the orbit of K in the first action consists of the left cosets of K in G, and the stabilizer of K is K itself. Thus 1. 2. 3 gives Lagrange's theorem that $|K|$ divides $|G|$, the quotient being the index $|G : K|$ or the number of distinct (left) cosets.

As another example, consider the second action in the case where $K = \{k\}$; then the orbit of K is the conjugacy class containing k. Moreover, the stabilizer G_k is just the centralizer $C(k)$ so that, by 1. 2. 4:

9

1.4.1   The number of conjugates of k is $|G : C(k)|$.

In particular, each element in the centre $Z(G)$ is its own conjugacy class. The partition of G induced by this action gives rise to the <u>class equation</u>

1.4.2 $|G| = |Z(G)| + \sum |G : C(g)|$,

where the summation ranges over a complete set of nonconjugate g not in $Z(G)$. It follows directly that a nontrivial p-group has a nontrivial centre. For if g is in $G - Z(G)$, then $C(g)$ is a proper subgroup of G and thus $|G : C(g)|$ is a positive power of p; thus p divides the summation and hence also $|Z(G)|$.

As a final example, we begin the proof of Sylow's theorem. Suppose p^r is the highest power of the prime p which divides $|G|$. Consider all subsets of G with cardinality p^r; there are $\binom{|G|}{p^r}$ of them, and this binomial coefficient is easily seen to be prime to p. Thus there must be at least one orbit (in the action $K \mapsto gK$) whose length m is prime to p, and the stabilizer G_L of a subset L in this orbit has order $|G|/m$, which is divisible by p^r. But if $l \in L$, then $G_L(l) \subseteq L$, and $|G_L| = |G_L(l)| \le |L| = p^r$, so that $|G_L| = p^r$. Thus we have produced a subgroup $(G_L$, in fact) of order p^r. The other parts of the theorem can be proved by similar arguments; see 1.8.

Next, we observe the relationship between subgroups of G and transitive permutation representations of G, namely: G has a transitive permutation representation of degree n if and only if G has a subgroup of index n. To see this, first let (G, X) be transitive, of degree n, so that (by 1.3.2) $|G : G_x| = |X| = n$ and G_x is a subgroup of index n. Conversely, if H is a subgroup of index n and $X = \{H, g_2H, \ldots, g_nH\}$ is the set of left cosets of H in G, then G acts transitively on X by defining $g(g_iH)$ to be the coset $(gg_i)H$. As a trivial application of this relationship, we see that S_5 has no transitive permutation representation of degree 4.

10

1.5   Extensions of multiply transitive groups

Apart from the symmetric and alternating groups, it is not easy to find k-transitive groups for values of k larger than 3; in fact, only two such 4-transitive groups, and two 5-transitive groups, are known. Since these groups are so rare, we shall devote some time to their construction. The basic idea is that if we are given a k-transitive group, then we may be able to add one new point to the underlying set and one new permutation to the group, in such a way that the old permutations and the new one generate a group acting (k+1)-transitively.

1.5.1 Definition. Let (G, X) be a transitive permutation group, and $X^+ = X \cup \{*\}$, where $*$ is not a member of X. We say that (G^+, X^+) is a <u>one-point extension</u> of (G, X) if G^+ is transitive on X^+ and the stabilizer $(G^+)_*$ is just G. (G is considered to act on X^+ in such a way that all its elements fix $*$.)

It follows from 1.3.6 that if G is k-transitive on X, then G^+ is (k+1)-transitive on X^+. Moreover, if G is sharply k-transitive on X, then G^+ is sharply (k+1)-transitive on X^+.

An obvious example is that S_{n+1} is a one-point extension of S_n, the new point $*$ being the symbol $n + 1$. In order to use the construction to define multiply transitive groups, we try to find a permutation h of X^+ such that $G^+ = \langle G, h \rangle$ (the group generated by G and h) has the right properties. Clearly, we must ensure that $h(*)$ is in X, so that G^+ will be transitive, but this alone is not enough. If h is chosen carelessly, G^+ will be too big - that is, we might obtain $\mathrm{Alt}(X^+)$ or $\mathrm{Sym}(X^+)$. For example, if we consider D_8 acting on the corners of a square, and let $* = 5$, then adding $h = (12345)$ gives a group $D_8^+ = S_5$.

In order to see what additional conditions h must satisfy, we examine the situation when an extension is known to exist. Suppose that H acts on X^+ in such a way that the stabilizer $H_* = G$ is multiply transitive on X, and let x, y be any two distinct points of X. Since H is (at least) 3-transitive, there is some h in H such that h switches $*$ and x, and fixes y. Also, since G is (at least) 2-transitive there is some g in G which switches x and y. It follows that both $(gh)^3$ and

h^2 fix *, and so they belong to G; also, if $f \in G_x$ then hfh fixes *
and x, so that $hG_x h = G_x$. We shall show that the existence of h and
g, satisfying these conditions, is also sufficient for the existence of a
one-point extension.

1.5.2 **Theorem.** <u>Let</u> (G, X) <u>be a k-transitive group with</u> $k \geq 2$,
<u>and let</u> $X^+ = X \cup \{*\}$. <u>Suppose that we can find a permutation</u> h <u>of</u> X^+
<u>and an element</u> g <u>of</u> G <u>such that</u>
  (i)   h <u>switches</u> * <u>and some</u> x <u>in</u> X, <u>and</u> h <u>fixes some point</u> y;
  (ii)   g <u>switches</u> x <u>and</u> y;
  (iii) $(gh)^3$ <u>and</u> h^2 <u>are in</u> G;
  (iv) $hG_x h = G_x$.
<u>Then the group</u> $G^+ = \langle G, h \rangle$ <u>acts on</u> X^+ <u>as a one-point extension of</u>
(G, X).

Proof. Since G is multiply transitive and $g \notin G_x$, we know from
1.3.9 that $G = G_x \cup G_x g G_x$. We shall show that the conditions imply that
$\langle G, h \rangle = G \cup GhG$; the result then follows, since nothing in GhG can fix
*, and so $(G^+)_* = G$.

It is sufficient to show that $G \cup GhG$ is a group. To show that
$G \cup GhG$ is closed under composition, we need only check that hGh is
a subset of $G \cup GhG$, since then we have

$$GhG \cdot GhG = G \cdot hGh \cdot G \subseteq G \cup GhG.$$

Now, h^2 fixes x, and $h^2 \in G$, so $h^2 \in G_x$; thus, by (iv) $hG_x = G_x h$.
Also $(gh)^3 \in G$, so that hgh belongs to

$$(ghg)^{-1}G = g^{-1}h^{-1}G = g^{-1}hG.$$

These remarks justify the following calculation:

$$
\begin{aligned}
hGh &= h(G_x \cup G_x g G_x)h \\
    &= hG_x h \cup hG_x g G_x h \\
    &= G_x \cup G_x \cdot hgh \cdot G_x \\
    &\subseteq G \cup G_x \cdot g^{-1}hG \cdot G_x \\
    &\subseteq G \cup GhG,
\end{aligned}
$$

as required. Thus $G \cup GhG = \langle G, h \rangle$. //

In practice g and h are often chosen so that both $(gh)^3$ and h^2 are the identity, and then condition (iii) of Theorem 1.5.2 is immediately satisfied.

We shall meet decidedly nontrivial applications of this theorem in Section 1.9 and in Chapters 2 and 3. For the moment, we consider only a result which could be obtained more simply. Suppose G is S_{n-1} acting on $X = \{1, 2, \ldots, n - 1\}$. Take * to be the symbol n, $h = (n - 1 \ n)$, and $g = (n - 2 \ n - 1)$. The conditions (iii) and (iv) are easily checked, so that G^+ is an n-transitive group and must be S_n. This gives us an inductive proof that S_n is generated by the transpositions $(i \ i + 1)$, for $1 \le i \le n - 1$. This could be claimed as one of the oldest results in permutation-group theory, since it is the essence of the art of change-ringing, and in that context it has been known (for small values of n) since the early seventeenth century.

1.6 Primitivity

1.6.1 Definition. Let (G, X) be a transitive permutation group, and suppose R is an equivalence relation on X. R is said to be G-admissible if $(x, y) \in R$ implies $(gx, gy) \in R$ for all $g \in G$.

The universal relation $R = X \times X$ and the equality relation, or diagonal Δ ($(x, y) \in \Delta$ if and only if $x = y$), are G-admissible, for any G. We shall say that these examples are trivial. A nontrivial example is provided by the action of the group of isometries on the corners of a regular 2m-gon in the plane. The relation of being on the same diagonal line is admissible for this action, and it is nontrivial.

1.6.2 Definition. The transitive permutation group (G, X) is primitive if there are no nontrivial G-admissible equivalence relations on X. Otherwise, it is imprimitive.

The aforementioned example shows that the dihedral group D_{4m} is imprimitive in its action on the corners of a regular 2m-gon. However, when the polygon has a prime number of sides, the group of isometries acts primitively.

13

Suppose that R is admissible for the action of G on X, and let $\bar{x}$ denote the R-equivalence class of x. We have a permutation representation $g \mapsto \bar{g}$ of G on the set X/R of equivalence classes, by defining $\bar{g}$ in Sym(X/R) as follows:

$$\bar{g}(\bar{x}) = \overline{gx} \qquad (g \in G, \ x \in X).$$

It is easy to check that $\bar{g}$ is well-defined. We note that, in general, the representation $g \mapsto \bar{g}$ will not be faithful: indeed, when $R = X \times X$ the set X/R has only one element and each $\bar{g}$ is the identity. In the case of a nontrivial relation R the set X/R has more than one element and is different from X. Thus the action of G on X has been reduced to its representation on X/R together with an action within each equivalence class. This provides the motivation for the study of primitive groups.

Although the proof that a given group is primitive may be difficult, there is one extremely useful result.

1.6.3 Theorem. _If_ (G, X) _is_ k-transitive ($k \geq 2$), _then it is primitive._

Proof. Suppose that R is G-admissible, and $R \neq \Delta$, so that we can find x and y such that $x \neq y$ and $(x, y) \in R$. For any z in X there is some g in G for which $g(x) = x$ and $g(y) = z$. Thus $(x, z) \in R$, and since R is an equivalence relation, $R = X \times X$. $/\!/$

The notion of primitivity has important consequences for the subgroup structure of a transitive group G. The rest of this section is devoted to these consequences.

1.6.4 Lemma. _Suppose that_ (G, X) _is a transitive permutation group, and_ H _is a subgroup of_ G _such that, for some_ $x \in X$, $G_x \leq H$. _Then_

$$R = \{ (g(x), \ gh(x)) \, | \, g \in G, \ h \in H \}$$

_is a_ G-admissible equivalence relation. Furthermore,

$$R = \Delta \quad \Longleftrightarrow \quad H = G_x.$$

$$R = X \times X \Longleftrightarrow H = G.$$

Proof. It is easy to see that R is G-admissible and that R is reflexive and symmetric. To show that R is transitive, suppose $(s, t) \in R$ and $(t, u) \in R$, so that

$$s = g_1 x, \quad t = g_1 h_1 x, \quad t = g_2 x, \quad u = g_2 h_2 x.$$

Then $g_2^{-1} g_1 h_1$ fixes x, and consequently belongs to H; thus $g_2^{-1} g_1$ and its inverse $g_1^{-1} g_2$ belong to H. Putting $g^* = g_1$ and $h^* = g_1^{-1} g_2 h_2$, we have $s = g^* x$, $u = g^* h^* x$, and so $(s, u) \in R$.

Suppose $R = \Delta$; then $s = gx$ and $t = ghx$ must imply $s = t$, for all $g \in G$ and $h \in H$. Taking $g = 1$ we see that h fixes x and so $H = G_x$. The converse is equally straightforward.

Suppose $R = X \times X$; then for any $k \in G$, $(x, kx) \in R$ and so there must be elements $g \in G$ and $h \in H$ such that $x = gx$ and $kx = ghx$. Thus $g \in G_x \leq H$ and $h^{-1} g^{-1} k \in G_x \leq H$, which implies that $k \in H$, $H = G$. The converse follows from the transitivity of G. //

1.6.5 Theorem. <u>The transitive permutation group</u> (G, X) <u>is primitive if and only if the stabilizer</u> G_x <u>is a maximal subgroup of</u> G. (<u>That is,</u> $G_x \leq H \leq G$ <u>implies</u> $H = G_x$ <u>or</u> $H = G$.)

Proof. The lemma shows that if G is primitive, then G_x is maximal. Conversely, suppose G_x is maximal and R is a G-admissible equivalence relation. Define

$$H = \{h \in G \mid (x, h(x)) \in R \}.$$

Then it follows that H is a subgroup of G containing G_x. Thus, either $H = G_x$ or $H = G$. If $H = G_x$, then x is R-equivalent only to itself, and since R is G-admissible and G is transitive, every equivalence class is a singleton: thus $R = \Delta$. If $H = G$, then the transitivity of G implies that X is an R-equivalence class, and so $R = X \times X$. Thus G is primitive. //

15

Another way of expressing the arguments in the preceding lemma and theorem is as follows. Given the transitive group (G, X) and x in X, there is a one-to-one correspondence between the lattice of subgroups of G which contain G_x and the lattice of G-admissible equivalence relations on X.

We now turn to the normal subgroup structure of primitive groups. We write $N \trianglelefteq G$ for the statement that N is a normal subgroup of G, possibly equal to G.

1.6.6 Theorem. <u>If</u> (G, X) <u>is primitive and</u> $N \trianglelefteq G$, $N \neq 1$, <u>then</u> N <u>is transitive on</u> X.

Proof. Define a relation R on X by $(x, y) \in R$ if and only if $y = n(x)$ for some $n \in N$. R is an equivalence relation, since N is a subgroup. Further, it is G-admissible, since the normality of N justifies the following argument:

$$(x, y) \in R \Rightarrow y = n(x)$$
$$\Rightarrow g(y) = gn(x) = n'g(x)$$
$$\Rightarrow (gx, gy) \in R.$$

Now $N \neq 1$ implies that $R \neq \Delta$, and so, since G is primitive, we must have $R = X \times X$. Thus N is transitive on X. //

The much-quoted example of D_8 acting on the corners of a square illustrates the two preceding theorems. In this example, the group acts imprimitively; the stabilizer of the corner 4 is $\{1, (13)\}$, which is not maximal as it is contained in the proper subgroup $\{1, (13), (24), (13)(24)\}$. Also, the normal subgroup $\{1, (13)(24)\}$ is not transitive.

Recall that a group is <u>simple</u> if it has no proper non-identity normal subgroups. Simple groups are of great importance in pure group theory, and it is often possible to prove that a group is simple by constructing a set on which it acts primitively. The relationship between the two concepts is indicated by the preceding theorem and the following one.

1.6.7 Theorem. <u>If</u> (G, X) <u>is primitive and</u> G_x <u>is simple, then</u> <u>either</u>

16

(i)   G is simple, or

(ii)   G has a normal subgroup N which acts regularly on X.

Proof.   Suppose that (i) is false, so that G has a proper normal subgroup N, $N \neq 1$. Given x in X, consider $N \cap G_x$. It is normal in G_x, and since G_x is simple, it must be either G_x itself or 1. Now $N \cap G_x = G_x$ means that $G_x \leq N$, and we must have $G_x < N$ since N is transitive and G_x is not. Thus, by 1.6.5, $N = G$, which contradicts the assumption that N is proper. It follows that $N \cap G_x = 1$, whence N acts regularly. //

1.7   Regular normal subgroups

The result 1.6.7 indicates the importance of regular normal subgroups, which form the subject of the present section. It is also possible to view the contents of this section as the first of our attempts to construct permutation groups by investing the set X of permuted objects with a structure which must be preserved; in this case the structure will be that of an abstract group. In order to state the connection between these two topics we must first define equivalence of permutation groups.

1.7.1 Definition.   The permutation groups (G, X) and (G', X') are equivalent if there is an isomorphism $g \mapsto g'$ of G and G', and a one-to-one correspondence $\beta : X \to X'$ such that

$$g'(\beta x) = \beta(gx)$$

for all $g \in G$ and $x \in X$.

If the permutation group (G, X) and the one-to-one correspondence $\beta : X \to X'$ are given, then the equation $g'(\beta x) = \beta(gx)$ defines an isomorphism $g \mapsto g'$. For example, suppose $X = \{1, 2, 3, 4\}$ is the set of corners of a square, and $X' = \{i, -1, -i, 1\}$, where the objects are to be regarded as complex numbers. Let $\beta : X \to X'$ be the correspondence given by $\beta(x) = i^x$. The action of D_8 on X induces an equivalent action on X'; for instance, if $g = (1234)$, $h = (13)$, then $g' = (i, -1, -i, 1)$ and $h' = (i, -i)$; that is, $g'(z) = iz$ and $h'(z) = \bar{z}$.

1.7.2 **Lemma.** Let (G, X) be a transitive permutation group and suppose N is a normal subgroup of G, acting regularly on X. Given $x \in X$, there is an action of G_x on $N^* = N - \{1\}$, equivalent to its action on $X - \{x\}$. Furthermore, the elements of G_x act as automorphisms of N.

Proof. Define $\beta : X - \{x\} \to N^*$ by $\beta(y) = n_y$, where n_y is the unique element of N which takes x to y. Given g in G_x, the corresponding permutation g' of N^* is defined by

$$g'(n_y) = n_{gy} .$$

This action, by definition, is equivalent to the action of G_x on $X - \{x\}$. Now gn_yg^{-1} is an element of N taking x to gy, and since N is regular, $n_{gy} = gn_yg^{-1}$. Thus, each g in G_x acts by conjugation, $g'(n) = gng^{-1}$, on N^*. If we extend the action to N in the obvious way $(g'(1) = 1)$, then we have an automorphism of N. //

Thus if a transitive permutation group (G, X) has a regular normal subgroup N, N must admit a group of automorphisms equivalent to the action of the stabilizer G_x on $X - \{x\}$. We show that few groups admit multiply transitive groups of automorphisms, thus restricting the possibilities for N. Note that any automorphism of a group H fixes the identity, so we are only concerned with the action on $H^* = H - \{1\}$. Let A be a group of automorphisms of H, and Z_p the cyclic group of order p.

1.7.3 **Theorem.** If A is transitive on H^*, then $H \approx (Z_p)^n$, for some prime p.

Proof. Each $a \in A$ preserves the order of an element of H, whence every non-identity element of H must have the same order, a prime p. Thus H is a p-group and its centre Z(H) is nontrivial (as we saw in Section 1.4). But each $a \in A$ preserves Z(H) and so, since A is transitive, H = Z(H); that is, H is abelian. The structure theorem for finite abelian groups now gives $H \approx (Z_p)^n$. //

1.7.4 Theorem. If A is 2-transitive on H*, then $H \approx (\mathbb{Z}_2)^n$ or $H \approx \mathbb{Z}_3$.

Proof. Define a relation R on H* by $(h, k) \in R$ if and only if either $h = k^{-1}$ or $h = k$; then R is an A-admissible equivalence relation on H*. Since A is primitive (1.6.3) it follows that either $R = \Delta$ or $R = H^* \times H^*$. If $R = \Delta$, then $(h, h^{-1}) = (h, h)$ for all $h \in H^*$, so $h = h^{-1}$ and $H \approx (\mathbb{Z}_2)^n$. If $R = H^* \times H^*$, then $\{h, h^{-1}\} = H^*$ and $H \approx \mathbb{Z}_3$. //

1.7.5 Theorem. If A is 3-transitive on H*, then $H \approx (\mathbb{Z}_2)^2$.

Proof. We may suppose $H \approx (\mathbb{Z}_2)^n$, $n \geq 2$, by virtue of the previous result and the exclusion of $\mathbb{Z}_2$ and $\mathbb{Z}_3$ because they are too small. Thus H contains a four-group $\{1, h, k, hk\}$. Now define an A_h-admissible equivalence relation R on $H^* - \{h\}$ by $(h_1, h_2) \in R$ if and only if either $h_1 = hh_2$ or $h_1 = h_2$. Now A_h is 2-transitive, and thus primitive, on $H^* - \{h\}$, and $\{k, hk\}$ is an R-equivalence class. Thus $R \neq \Delta$, so that $H^* - \{h\} = \{k, hk\}$ and $H \approx (\mathbb{Z}_2)^2$. //

1.7.6 Theorem. If (G, X) is k-transitive and has a regular normal subgroup N, then

    (i) $k = 2 \Rightarrow N \approx (\mathbb{Z}_p)^n$, $|X| = p^n$, p prime.

    (ii) $k = 3 \Rightarrow N \approx (\mathbb{Z}_2)^n$ or $\mathbb{Z}_3$, $|X| = 2^n$ or 3.

    (iii) $k = 4 \Rightarrow N \approx (\mathbb{Z}_2)^2$, $|X| = 4$.

    (iv) $k \neq 5$.

Proof. If (G, X) is k-transitive, $(G_x, X - \{x\})$ is (k-1)-transitive and so, by 1.7.2, N admits a (k-1)-transitive group of automorphisms. 1.7.3, 1.7.4, and 1.7.5 yield the result. //

As an application we prove the simplicity of the alternating groups.

1.7.7 Lemma. A_5 is simple.

Proof. The conjugacy classes of A_5 are those of S_5 which consist of even permutations, except that, since (12345) and (21345)

19

are conjugates in S_5 only by odd permutations, this class splits in A_5 into two classes of twelve permutations each. The classes are thus:

Representative:	1	(12)(34)	(123)	(12345)	(21345)
Number:	1	15	20	12	12

A normal subgroup is a union of classes including the identity class - but no such union in this case has cardinal number dividing 60. //

1.7.8 Theorem. A_n <u>is simple</u>, $n \geq 5$.

Proof. We use induction on n, the initial step being given by 1.7.7. For $n > 5$, the stabilizer $(A_n)_n$ is A_{n-1}, which we assume to be simple. Now A_n is $(n-2)$-transitive, by 1.3.8, and so has no regular normal subgroup, by 1.7.6. Also, A_n is primitive, by 1.6.3. Thus A_n is simple, by 1.6.7. //

1.8   Project: Proof of Sylow's theorem

Let p^r be the highest power of the prime p which divides $|G|$. We have seen that G contains a subgroup of order p^r - that is, a <u>Sylow p-subgroup.</u> Now complete the proof of Sylow's theorem, as follows.

1.8.1   Use the isomorphism theorem $(S \leq K, T \trianglelefteq K \Rightarrow ST \leq K$ and $S/(S \cap T) \approx (ST)/T)$ to show that a Sylow p-subgroup P of G is the only Sylow p-subgroup of its normalizer $N(P) = \{n \in G \mid nPn^{-1} = P\}$.

1.8.2   Let $X = \{P_1, P_2, \ldots, P_s\}$ be all the Sylow p-subgroups of G. Define a permutation representation of degree s of P_1 by $g \mapsto \bar{g}$, where $\bar{g}(P_i) = g^{-1}P_ig$, $1 \leq i \leq s$. Use 1.8.1 to show that P_1 is the only Sylow p-subgroup fixed by all $\bar{g}$ $(g \in P_1)$.

1.8.3   Now show that the number s of Sylow p-subgroups of G is congruent to 1 (modulo p).

1.8.4   Now suppose that $P_1, P_2, \ldots, P_t$ are conjugate in G, where t is maximal. Let P_1 act on these t objects, as above, and conclude that $t \equiv 1 \pmod{p}$. If $t < s$, then there is another system

P_{t+1}, P_{t+2}, ..., P_{t+k} of Sylow p-subgroups, conjugate in G; this gives two permutation representations of degree k, one of P_1 and one of P_{t+1}. Show that both $k \equiv 0$ and $k \equiv 1$ (mod p). Conclude that $t = s$, and any two Sylow p-subgroups of G are conjugate to each other.

1.8.5 Deduce that s divides $|G|$.

1.8.6 Let H be a subgroup of G whose order is a power of p. Define a permutation representation of H of degree s, and use this to find a P_k, $1 \le k \le s$, such that $H \le P_k$.

1.9   Project: Some multiply transitive groups

When the group G acts on a set Z it also permutes the subsets of Z, and so there are permutation representations of G induced in this way. (It is convenient to say that an element g of G fixes $Y \subseteq Z$ pointwise if $y \in Y$ implies $g(y) = y$, and setwise if $y \in Y$ implies $g(y) \in Y$.)

In this project we shall construct some multiply transitive groups starting from a representation of S_6. Let $Z = \{a, b, c, d, e, f\}$ and let X denote the set of 10 partitions of Z into two sets of three. Label the members of X as follows:

0	abc|def	5	ace|bdf
1	abd|cef	6	acf|bde
2	abe|cdf	7	ade|bcf
3	abf|cde	8	adf|bce
4	acd|bef	9	aef|bcd

1.9.1 Let $g \mapsto \hat{g}$ denote the representation of $S_6 = Sym(Z)$ as permutations of X. Check that this is a faithful representation and that $\hat{S}_6$ acts transitively on X. Find the orders of

(i)   the pointwise stabilizer of abc in S_6;

(ii)   the setwise stabilizer of abc in S_6;

(iii)   the stabilizer of 0 in $\hat{S}_6$.

1.9.2 By considering (abc)^ and (def)^, show that $\hat{S}_6$ is

2-transitive on **X**.

1. 9. 3   How many elements of $\hat{S}_6$ fix both 0 and 1? Find them. Deduce that $\hat{S}_6$ is not 3-transitive on **X**.

1. 9. 4   Let $H = \hat{A}_6$ be the group of permutations of **X** induced by the even permutations of **Z**. Verify that **H** is 2-transitive on **X** and that H_{01} is a cyclic group of order 4 generated by
$\theta = (afbe)(cd)\hat{} = (2934)(5876)$.

1. 9. 5   Show that H_0 is generated by θ, $\phi_1 = (abc)\hat{}$ and $\phi_2 = (def)\hat{}$. Use the fact that **H** is primitive on **X** to deduce that **H** is generated by θ, ϕ_1, ϕ_2, and any element ψ of $H - H_0$.

1. 9. 6   Take ψ to be $(ab)(cd)\hat{} = (01)(49)(56)(78)$, so that $H = \langle \theta, \phi_1, \phi_2, \psi \rangle$. Define a new permutation, not in **H**, by $\lambda = (2735)(4698)$. Show that the conjugate of each generator of **H** by λ is an element of **H** and that $\lambda^2 \in H$. Deduce that there are just two cosets of **H** in $\langle H, \lambda \rangle$.

1. 9. 7   Define M_{10} (the <u>Mathieu group</u> on 10 symbols) to be $\langle H, \lambda \rangle$. Verify that M_{10} is sharply 3-transitive on **X**.

1. 9. 8   In 1.5.2, let $G = M_{10}$, $* = T$, and $h = (0T)(47)(59)(68)$, $g = \psi$. Show that $G_0 = \langle \theta, \phi_1, \phi_2, \lambda \rangle$ and verify that the conditions of the theorem hold, so that $M_{11} = \langle M_{10}, h \rangle$ is sharply 4-transitive on $X' = \{0, 1, \ldots, 9, T\}$.

1. 9. 9*   Now, let $G' = M_{11}$, $*' = E$, and $g' = h$. Find a permutation h' of $\{0, 1, \ldots, 9, T, E\}$ such that $M_{12} = \langle M_{11}, h' \rangle$ is sharply 5-transitive. (Hint: Take $h'(E) = T$ and show that h' must fix 0 and have the same cycle shape as h.) M_{11} and M_{12} are the <u>Mathieu groups</u> of degree 11 and 12 respectively.

1. 9. 10   Suppose **Z** is replaced by a set with 8, 10, ... members; at what stage does the preceding construction break down?

1.9.11   Is M_{10} simple? Assume that M_{11} is simple, and show that M_{12} is simple. (In 3.7.9 we shall see that M_{11} is indeed simple.)

NOTES AND REFERENCES FOR CHAPTER 1

The study of groups began with the work of Lagrange, Galois, Cauchy, and others, on groups of substitutions (permutations). The notions of transitivity and primitivity emerged from the attempts of nineteenth-century mathematicians to classify the permutation groups of small degree.

Most books on group theory contain an introduction to permutation groups, but there are few books wholly devoted to the subject. The two standard works are by Wielandt [3] and Passman [1].

The proof of Sylow's theorem in Sections 1.4 and 1.8 may be traced back to G. A. Miller; its present form is due to Wielandt [2]. The Mathieu groups were discovered by Mathieu in 1861, and investigated by Witt [4].

1.   Passman, D. S. Permutation Groups. (Benjamin, 1968).
2.   Wielandt, H. Ein Beweis für die Existenz der Sylowgruppen. Arch. Math. , 10 (1959), 401-2.
3.   Wielandt, H. Finite Permutation Groups. (Academic Press, 1964.)
4.   Witt, E. Die 5-fach transitiven Gruppen von Mathieu. Abh. Math. Sem. Univ. Hamburg, 12 (1938), 256-64.

2 · Finite Geometries

'It is my experience that proofs involving matrices can be shortened by half if one throws the matrices out. Sometimes it cannot be done; a determinant may have to be computed. '
E. Artin, in the introductory chapter of Geometric algebra, 1957.

2.1   Introduction

If we are given a finite set X, we obtain the permutation groups Sym(X) and Alt(X) without further constructions. In order to derive more interesting groups, we may impose some structure on X and investigate the subgroup of Sym(X) containing those permutations which preserve the structure.

We begin with an illustrative example. A finite projective plane is a finite set X, together with a family $\mathcal{L}$ of subsets of X, satisfying three axioms:

PP1.   Each pair of distinct elements of X belongs to exactly one set l in $\mathcal{L}$.

PP2.   The intersection of each pair of distinct elements of $\mathcal{L}$ is a single member of X.

PP3.   There are at least four members of X having the property that no three of them belong to a single member of $\mathcal{L}$.

It is customary to refer to the elements of X as points, and the elements of $\mathcal{L}$ as lines. It is easily verified that if X = {1, 2, ..., 7}, then the family $\mathcal{L}$ of subsets 124, 235, 346, 457, 561, 672, 713 satisfies the axioms; a pictorial representation of this 'seven-point plane' is shown in Fig. 1.

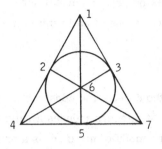

Fig. 1

A permutation which 'preserves the structure' of a projective plane is simply a permutation π of X with the property that $\pi(l)$ is in $\mathcal{L}$ whenever l is. This concept might be denoted by the general term 'automorphism', or the special name 'collineation' (see Section 2.6). It is clear that the set of automorphisms forms a group, under the usual composition operation for permutations. We can use the ideas of Chapter 1 to investigate this group in the case of the seven-point plane.

The group G of automorphisms of the seven-point plane is transitive, since it contains (1234567); it is 2-transitive, since the stabilizer G_1 contains (276)(435) and (2347)(56). However, an automorphism which fixes 1 and 2 must fix 4 also; thus G is not 3-transitive. It is easy to see that G_{12} consists of four permutations only:

$$G_{12} = \{\text{identity, } (37)(56), (36)(57), (35)(67)\},$$

so that the order of G is $7.6.4 = 168$. It is fairly easy to determine the conjugacy classes in G, along the following lines. An automorphism like (276)(435) commutes only with itself, its inverse, and the identity, so that its centralizer has order 3; by 1.4.1, its conjugacy class has order 56. A 7-cycle is centralized by its seven powers, and so its conjugacy class has order 24; however, there are 8 Sylow-7-subgroups, which implies 48 7-cycles in all - two classes of 24. A complete set of representatives, and the sizes of the respective classes, is as follows:

Identity	(1234567)	(1765432)	(276)(435)	(24)(56)	(2347)(56)
1	24	24	56	21	42

We see that no union of classes (including the identity) has cardinality dividing 168, so that G has no proper normal subgroups. Later in this chapter we shall see that G is one of an infinite family of simple groups.

2.2 Finite fields

We shall sketch without proof the construction and properties of finite fields. The ring $\mathbb{Z}$ of integers induces a natural ring structure on $\mathbb{Z}_n = \mathbb{Z}/n\mathbb{Z}$, the integers modulo n; if n is a prime p then $\mathbb{Z}_p$ is in fact a field under this structure. Similarly, the set $(\mathbb{Z}_n)^r$ of sequences $(a_0, a_1, \ldots, a_{r-1})$, with $a_i \in \mathbb{Z}_n$, has an additive structure in which addition is performed coordinate-wise, and again when n is a prime, this set may be given a multiplicative structure which makes it a field. The field is obtained in the following way:

(a)   identify the sequence $(a_0, a_1, \ldots, a_{r-1})$ with the polynomial $a_0 + a_1 t + \ldots + a_{r-1} t^{r-1}$ in the ring of polynomials $\mathbb{Z}_p[t]$;

(b)   choose a polynomial f(t) of degree r which is irreducible (has no proper factors) in $\mathbb{Z}_p[t]$;

(c)   define multiplication of two sequences by multiplying the corresponding polynomials in $\mathbb{Z}_p[t]$ and then reducing modulo f(t).

It is always possible to choose f(t) in such a way that the non-zero elements of the field are just the powers

$$t, t^2, \ldots, t^{p^r - 1},$$

the last of these being the multiplicative identity 1. The field constructed in this way is sometimes called the Galois field with p^r elements, and it is denoted by $GF(p^r)$.

For example, in order to construct $GF(3^2)$ we note that $t^2 + 2t + 2$ is irreducible over $\mathbb{Z}_3 = \{0, 1, 2\}$; thus the elements of $GF(3^2)$ may be listed as follows:

	0	t	t^2	t^3	t^4	t^5	t^6	t^7	t^8
reduction:	0	t	1+t	1+2t	2	2t	2+2t	2+t	1
sequence:	(0,0)	(0,1)	(1,1)	(1,2)	(2,0)	(0,2)	(2,2)	(2,1)	(1,0)

We summarize the relevant properties of finite fields.

2.2.1 (i) There is a finite field with n elements if and only if n is a prime power, $n = q = p^r$.

(ii) If F is a finite field with q elements, then F is isomorphic with the Galois field $GF(q)$; in particular, the structure of the field does not depend on the choice of irreducible polynomial $f(t)$.

(iii) The multiplicative group of $GF(p^r)$ is a cyclic group of order $p^r - 1$. A generator of this group is called a _primitive_ element of the field.

(iv) The group of field automorphisms of $GF(p^r)$ is a cyclic group of order r generated by the automorphism $x \mapsto x^p$.

The finite fields yield families of 2-transitive and 3-transitive groups.

2.2.2 Theorem. <u>Let F be a finite field. The set A of affine transformations of F,</u>

$$u \mapsto au + b, \quad a, b \in F, \ a \neq 0,$$

<u>is a group which acts sharply 2-transitively on F.</u>

Proof. That A is a group (under composition) is immediate. It acts sharply 2-transitively, since given x, y in F the equations $x = a0 + b$, $y = a1 + b$ have the unique solution $a = y - x$, $b = x$. In other words, there is a unique affine transformation taking $0, 1$ to x, y. //

2.2.3 Theorem. <u>Let L denote the set $F \cup \{\infty\}$, where F is a finite field and ∞ is a new symbol, not in F. Then there is a group P which acts sharply 3-transitively on L.</u>

Proof. We shall use Theorem 1.5.2 to construct a one-point extension of (A, F). In the notation of that theorem, take $G = A$, $X = F$, $* = \infty$ and $x = 0$. The extending permutation h is defined by

$$h(\infty) = 0, \quad h(0) = \infty, \quad h(u) = u^{-1} \quad (u \in F - \{0\}),$$

and g is the affine transformation $u \mapsto 1 - u$. It is easy to check that the conditions of 1.5.2 are satisfied. Since $gh(u) = 1 - u^{-1}$, it follows that $(gh)^3$ is the identity, as is h^2. The stabilizer G_0 is the cyclic group generated by the transformation $u \mapsto tu$, where t is a primitive element of F, and conjugating by h gives the transformation $u \mapsto t^{-1}u$, whence $hG_0 h = hG_0 h^{-1} = G_0$. Thus we may conclude that $P = \langle A, h \rangle$ acts sharply 3-transitively on L. $/\!/$

2.3 Finite vector spaces

We shall write $V = V(n, q)$ for a vector space of dimension n over $F = GF(q)$. V has q^n elements. Our aim is to use standard techniques of linear algebra to define structures of geometrical and combinatorial interest; these structures, in turn, give rise to interesting groups.

2.3.1 Definition. A linear automorphism of $V(n, q)$ is a permutation l of V satisfying

$l(x + y) = l(x) + l(y)$ for all $x, y \in V$,

$l(\lambda x) = \lambda l(x)$ for all $\lambda \in F$ and $x \in V$.

Choosing a basis $\{e_1, e_2, \ldots, e_n\}$ for V, we can define the coordinates of a point x in V to be the unique ordered n-tuple $(x_1, x_2, \ldots, x_n)$ of elements of F such that $x = x_1 e_1 + \ldots + x_n e_n$. A linear automorphism of V is specified by its action on the basis, and we may represent l by the $n \times n$ matrix over F, $L = (l_{ij})$, whose entries are given by

$$l(e_j) = \sum_{i=1}^{n} l_{ij} e_i .$$

The action of l on a point x corresponds to the action of the matrix L by left multiplication on the column vector whose entries are the co-ordinates of x.

If we had chosen a different basis $\{f_1, f_2, \ldots, f_n\}$ for V, then there is a unique linear automorphism k of V such that $k(e_i) = f_i$

$(1 \leq i \leq n)$. If $K = (k_{ij})$ is the matrix representing k with respect to the basis $\{e_1, \ldots, e_n\}$, then l is represented with respect to the new basis $\{f_1, \ldots, f_n\}$ by the matrix $K^{-1}LK$. Thus the determinant of a linear automorphism may be defined as the determinant of any representative matrix. We shall use the notation det l for this determinant.

It is easy to check that if the permutation l of V is a linear automorphism, then the inverse permutation l^{-1} is also a linear automorphism. Since we must have $(\det l)(\det l^{-1}) = \det(l \cdot l^{-1})$, and the determinant of the identity is 1, it follows that the determinant of a linear automorphism is non-zero. Clearly, the linear automorphisms of V form a group.

2.3.2 Definition. The <u>general linear group</u> GL(V) is the group of all linear automorphisms of V. The subgroup consisting of those linear automorphisms with determinant 1 is called the <u>special linear group</u> SL(V). When $V = V(n, q)$, we use the notations GL(n, q) and SL(n, q).

2.3.3 Lemma. <u>SL(V) is a normal subgroup of GL(V), and if</u> $V = V(n, q)$ <u>then the index</u> $|GL(V) : SL(V)|$ <u>is</u> $q - 1$.

Proof. The determinant function maps GL(V) into $F^* = F - \{0\}$, and elementary properties of the determinant imply that it is, in fact, a homomorphism onto the multiplicative group F^*. The kernel of this homomorphism is, by definition, SL(V). Hence SL(V) is a normal subgroup and its index in GL(V) is $|F^*| = q - 1$. $/\!/$

2.3.4 Theorem. <u>The orders of</u> GL(V) <u>and</u> SL(V) <u>are given by:</u>

$$|GL(n, q)| = (q^n - q^{n-1})(q^n - q^{n-2}) \ldots (q^n - 1)$$

$$= q^{n(n-1)/2} \prod_{i=1}^{n} (q^i - 1);$$

$$|SL(n, q)| = q^{n(n-1)/2} \prod_{i=2}^{n} (q^i - 1).$$

Proof. For each pair of ordered bases $\{e_1, \ldots, e_n\}$, $\{f_1, \ldots, f_n\}$ of $V(n, q)$ there is a unique linear automorphism l

taking e_i to f_i $(1 \le i \le n)$. Thus $|GL(n, q)|$ is equal to the number of ordered bases of $V(n, q)$. Now the first member of an ordered basis may be any element of V except 0, and so it may be chosen in $q^n - 1$ ways. The second element must not be linearly dependent on the first, and so it can be chosen in $q^n - q$ ways. Continuing in this way we obtain the formula for $|GL(n, q)|$. The formula for $|SL(n, q)|$ is obtained from Lemma 2.3.3. $/\!/$

Before we proceed to a detailed study of the structure of $GL(V)$ and $SL(V)$, we make some informal comments. Since every linear automorphism of V fixes 0, our interest in the transitivity properties of $GL(V)$ is necessarily confined to its action on $V^* = V - \{0\}$. Although $GL(V)$ is transitive on V^* (since any two elements of V^* may be chosen as the first members of two ordered bases), it is not generally 2-transitive, since there is no automorphism taking a linearly independent pair into a dependent pair.

When $V = V(n, p)$, p prime, the linear automorphisms of V are just the group automorphisms of the additive group of V, that is, $(\mathbb{Z}_p)^n$. This provides a geometrical interpretation of Theorem 1.7.4: over the field $GF(2)$ any two distinct points of V^* are linearly independent, and so we may expect to find 2-transitive groups of automorphisms.

2.4   The structure of $GL(V)$ and $SL(V)$

Let V denote a vector space over the field $F = GF(q)$. For the rest of this chapter we shall suppose, unless explicitly stated to the contrary, that $n = \dim V$ is at least 2. An m-dimensional subspace of $V(n, q)$ contains q^m points; when $m = n - 1$ a subspace is called a hyperplane and it contains q^{n-1} points. If u is a non-trivial linear functional on V, the set

$$U = \{x \in V \,|\, u(x) = 0\}$$

is a hyperplane, and $u(x) = 0$ is an equation for it. If $u(x) = 0$ and $t(x) = 0$ are both equations for the same hyperplane, then there is some λ in F^* such that $u(x) = \lambda t(x)$ for all x in V. Also, for any g in $GL(V)$, $g(U)$ is a hyperplane and an equation for it is $u'(x) = 0$, where u' is the composite $u \circ g^{-1}$.

2.4.1 Definition. A linear automorphism τ in GL(V) is a transvection, with direction $d \in V^*$, if τ fixes d and $\tau(x) - x$ is a scalar multiple of d for all x in V.

In other words, a transvection moves every point in a direction parallel to d. We note that the identity is a transvection (even when $n = 1$) for any direction d; also, a transvection with direction d is a transvection with direction λd, for all λ in F^*.

2.4.2 Theorem. If τ is a transvection with direction d, then there is a hyperplane containing d each of whose points is fixed by τ.

Proof. If τ is the identity, the result is obvious. If not, let $\tau(x) - x = u(x)d$ for each x in V. Since τ is linear, u is a non-trivial linear functional on V. Let U be the hyperplane with equation $u(x) = 0$; then $\tau(x) = x$ whenever x is in U, and $u(d) = 0$, as required. //

It follows that we can write any transvection in the form
$$\tau = \tau_{u, d}, \text{ where}$$
$$\tau_{u, d}(x) = x + u(x)d,$$

and u is linear functional, $0 \neq d \in V$, $u(d) = 0$.

2.4.3 Lemma. Given g in GL(V), and using the preceding notation for transvections, we have

(i) $\quad \tau_{u, d}\tau_{v, d} = \tau_{u+v, d}$,

(ii) $\quad g\tau_{u, d}g^{-1} = \tau_{u', d'}$,

where $u' = u \circ g^{-1}$ and $d' = g(d)$.

Proof. By direct calculation, using the formula. //

Most of the ensuing argument will be based on the formula for a transvection in the form $\tau_{u, d}$. However, at one point it will be convenient to use an explicit matrix representation for the case when $n = \dim V = 2$. In this case, if τ is a transvection with direction d, we may choose basic vectors $e_1 = d$ and $e_2 = z$ (not linearly dependent on d); then the matrix representing τ takes the form

$$\begin{bmatrix} 1 & \lambda \\ 0 & 1 \end{bmatrix} \qquad (\lambda \in F).$$

Let T denote the set of all transvections in GL(V), and let $T^* = T - \{1\}$; let T(d) denote the set of transvections with direction d. We shall analyse the structure of GL(V) and SL(V) in terms of T and T(d).

2.4.4 **Theorem.** Using the notation introduced above, we have:

(i) T^* is a complete conjugacy class in GL(V);

(ii) $T \subseteq SL(V)$;

(iii) if dim $V \geq 3$, then T^* is a complete conjugacy class in SL(V).

Proof. (i) The conjugate of a transvection is a transvection, by Lemma 2.4.3(ii). Conversely, suppose that $\tau = \tau_{u,d}$ and $\tau' = \tau_{u',d'}$ are two transvections. Choose bases $\{d, b, \ldots, e\}$ and $\{d', b', \ldots, e'\}$ for the hyperplanes U and U' with equations $u(x) = 0$ and $u'(x) = 0$ respectively; then select $f \notin U$, $f' \notin U'$ such that $u(f) = u'(f') = 1$. The sets $\{d, b, \ldots, e, f\}$ and $\{d', b', \ldots, e', f'\}$ are both bases for the whole space V, and consequently we can find $h \in GL(V)$ such that $h(d) = d', \ldots, h(f) = f'$. We claim that $\tau' = h\tau h^{-1}$. Both transvections have the same direction, since $d' = h(d)$. Also, $u'(x) = 0$, $(u \circ h^{-1})(x) = 0$ are both equations for the hyperplane U', so that there is some $\lambda \neq 0$ such that $u'(x) = \lambda(u \circ h^{-1})(x)$ for all x in V. Putting $x = f'$ we obtain $\lambda = 1$, so that τ' and $h\tau h^{-1}$ have the same formula.

(ii) Since all transvections are conjugate in GL(V), they all have the same determinant $\delta \neq 0$. Taking the determinant of the equation in 2.4.3(i), we obtain $\delta^2 = \delta$, and so $\delta = 1$.

(iii) Suppose that dim $V \geq 3$, and that τ and τ' are as in part (i) of the proof. Suppose also that the element h of GL(V) obtained in

(i) has determinant $\mu \neq 1$. Replacing the basis vector b by μb, and proceeding as before, we get h* in GL(V) with $\tau' = h^*\tau h^{*-1}$ and det h* $= 1$; that is, τ and τ' are conjugate in SL(V). (Note that two elements of the basis (d and f) are prescribed in (i), so that the condition dim $V \geq 3$ is necessary here.) //

2.4.5 Theorem. The set T(d) of transvections with direction d is an abelian normal subgroup of the stabilizer of d in the action of SL(V) on V*. The groups T(d) (d $\in$ V*) are all conjugate in SL(V).

Proof. We have just shown that $T \subseteq SL(V)$. T(d) is an abelian group, by 2.4.3(i), and if g is an element of SL(V) fixing d then $g\,T(d)g^{-1} = T(d)$, by 2.4.3(ii). Thus T(d) is an abelian normal subgroup of the stabilizer of d. If d_1 and d_2 are any two members of V*, there is an element of SL(V) taking d_1 to a scalar multiple of d_2, and $T(d_1)$ and $T(d_2)$ are conjugate under this element. //

We require a lemma. Suppose that $0 \neq y \in V$, and let $Y = \{\lambda y \,|\, \lambda \in F \}$. Denote the quotient vector space V/Y by $\overline{V}$; a typical element of $\overline{V}$ is the set

$$\overline{x} = \{x' \in V \,|\, x' - x \in Y \}.$$

2.4.6 Lemma. Let V, y, Y, and $\overline{V}$ be as above. Given g in SL(V) such that gy = y, put $\overline{g}(\overline{x}) = \overline{gx}$. Then
    (i) $\overline{g}$ is a well-defined element of SL($\overline{V}$);
    (ii)   if we are given a transvection σ in SL($\overline{V}$), there is a transvection τ in SL(V) such that $\overline{\tau} = \sigma$.

Proof. (i) It is easy to check that $\overline{g}$ is a well-defined linear automorphism of $\overline{V}$. Let $\{y, v_2, \ldots, v_n \}$ be a basis for V, and suppose that M is the corresponding matrix representing g; we are given that det M $= 1$. Since gy = y, the first column of M is the transpose of $(1, 0, \ldots, 0)$, so that det $\overline{M} = 1$, where $\overline{M}$ is obtained from M by deleting the first row and column. But $\overline{M}$ is the matrix representing $\overline{g}$ with respect to the basis $\{\overline{v}_2, \ldots, \overline{v}_n \}$ of $\overline{V}$; thus $\overline{g} \in SL(\overline{V})$.

(ii)   If a transvection σ in $SL(\overline{V})$ has direction $\overline{d} \in \overline{V}$, and its hyperplane has the equation $\overline{u}(\overline{x}) = \overline{0}$, then choose a representative d in V for $\overline{d}$, and define a linear functional u on V by

$$u(y) = 0, \quad u(x) = \overline{u}(\overline{x}) \quad (x \notin Y).$$

The transvection τ on V given by $\tau = \tau_{u,d}$ satisfies $\overline{\tau} = \sigma$, since

$$\begin{aligned}
\overline{\tau}(\overline{x}) &= \overline{\tau x} \\
&= \overline{x + u(x)d} \\
&= \overline{x} + u(x)\overline{d} \\
&= \overline{x} + \overline{u}(\overline{x})\overline{d} \\
&= \sigma(\overline{x}). \quad /\!/
\end{aligned}$$

2.4.7 Theorem.   <u>The set T of transvections generates</u> $SL(V)$.

Proof.   We have to show that every element of $SL(V)$ can be expressed as a product of transvections, and we shall do this by induction on $n = \dim V$. Clearly the result is true when $n = 1$, since then $T = SL(V) = 1$.

Suppose that $n \geq 2$, and g is in $SL(V)$. We consider first the case when g fixes some vector $y \neq 0$ in V. In this case, by Lemma 2.4.6(i), there is a corresponding $\overline{g}$ in $SL(\overline{V})$, and since $\dim \overline{V} = n - 1$ we may suppose inductively that $\overline{g}$ is a product of transvections:

$$\overline{g} = \sigma_1 \sigma_2 \cdots \sigma_k .$$

Now, using Lemma 2.4.6(ii), we can find transvections $\tau_1, \tau_2, \ldots, \tau_k$ in $SL(V)$ such that $\overline{\tau}_i = \sigma_i$ $(1 \leq i \leq k)$. Let $h = \tau_1 \tau_2 \cdots \tau_k$. We have

$$\begin{aligned}
\overline{hx} &= \overline{h}(\overline{x}) \\
&= \overline{\tau_1 \tau_2 \cdots \tau_k}(\overline{x}) \\
&= \overline{\tau}_1 \overline{\tau}_2 \cdots \overline{\tau}_k(\overline{x}) \\
&= \sigma_1 \sigma_2 \cdots \sigma_k(\overline{x}) \\
&= \overline{g}(\overline{x}) \\
&= \overline{gx} .
\end{aligned}$$

Thus hx - gx is a scalar multiple of y, and (since gy = y) so is g^{-1}hx - x, for all x in V. In other words, g^{-1}h is a transvection with direction y. Since h is a product of transvections, so is g. We have completed the induction step in the case when g has a fixed point y.

For the remaining cases, suppose first that g takes y to a linearly independent vector gy. Then y - gy and gy are linearly independent, and we can choose a linear functional t on V such that t(y - gy) = 0 and t(gy) = 1. Let θ be the transvection $\tau_{t, y-gy}$; then θg fixes y, and the preceding argument shows that θg is a product of transvections. Hence g is a product of transvections.

Suppose finally that g takes y to a linearly dependent vector gy. Now we choose a vector b, not linearly dependent on y, and a linear functional w such that w(gy) = 1 and w(b) = 0. Let $\phi = \tau_{w, b}$; then ϕg takes y to the linearly independent vector gy + b, and so we return to the case dealt with in the preceding paragraph. Thus, in all cases, g is a product of transvections, and the induction step is complete. //

2.4.8 Theorem. SL(n, q) <u>coincides with its commutator subgroup, provided</u> $n \geq 2$ <u>and</u> $(n, q) \neq (2, 2)$ <u>or</u> $(2, 3)$.

Proof. By 2.4.6, it is sufficient to show that any transvection can be written as a commutator $[a, b] = aba^{-1}b^{-1}$, a, b ϵ SL(n, q).

Suppose first that $n \geq 3$ and $\tau \epsilon T(d)$. Let $\sigma \epsilon T(d)$ $(\sigma \neq \tau^{-1})$, so that $\tau\sigma \epsilon T(d)$ also. By 2.4.4(iii), there is some g in SL(n, q) such that $\tau\sigma = g\sigma g^{-1}$: that is, τ is the commutator $[g, \sigma]$.

If n = 2, we use the matrix representation. Consider the identity

$$\begin{bmatrix} \alpha & 0 \\ 0 & \alpha^{-1} \end{bmatrix} \begin{bmatrix} 1 & \beta \\ 0 & 1 \end{bmatrix} \begin{bmatrix} \alpha^{-1} & 0 \\ 0 & \alpha \end{bmatrix} \begin{bmatrix} 1 & -\beta \\ 0 & 1 \end{bmatrix} = \begin{bmatrix} 1 & \beta(\alpha^2 - 1) \\ 0 & 1 \end{bmatrix},$$

which is valid over any field. If τ is a given transvection in SL(2, q), with matrix $\begin{bmatrix} 1 & \gamma \\ 0 & 1 \end{bmatrix}$, then the identity shows that we need only find α and β in GF(q) such that $\gamma = \beta(\alpha^2 - 1)$. Now, provided q $\neq$ 2, 3, we may take α to be a primitive element of GF(q) (so that $\alpha^2 \neq 1$), and set $\beta = \gamma(\alpha^2 - 1)^{-1}$. //

The preceding result is definitely not true in the exceptional cases SL(2, 2) and SL(2, 3). In both groups the commutator subgroup is a proper subgroup.

The last result in this section concerns the centre $Z(G)$ when G is GL(V) or SL(V). It is very easy to characterize the central elements of these groups, and to deduce, as a consequence, that most of the groups will have a non-trivial centre.

2.4.9 Theorem. <u>The centre of</u> GL(V) <u>consists of the scalar transformations</u> $x \mapsto \lambda x$ ($\lambda \in F^*$); <u>the centre of</u> SL(V) <u>consists of those scalar transformations for which</u> $\lambda^n = 1$, <u>where</u> $n = \dim V$.

Proof. It is clear that the scalar transformations belong to the centre of GL(V). Conversely, suppose g commutes with every element of GL(V); in particular, for each $x \in V^*$ it commutes with a transvection τ having direction x. Since $\tau = g\tau g^{-1}$, and $g\tau g^{-1}$ is a transvection with direction $g(x)$, there must be some λ_x in F^* such that $g(x) = \lambda_x x$. This holds for each x in V^*, and the linearity of g implies that

$$\lambda_{x+y}(x + y) = \lambda_x x + \lambda_y y \qquad (x, y \in V^*).$$

If x and y are linearly independent, we deduce that $\lambda_x = \lambda_y = \lambda_{x+y}$; if not, we choose z independent of both x and y, so that $\lambda_z = \lambda_x = \lambda_y$. Hence λ_x is constant and g is a scalar transformation.

The same argument holds in SL(V) (since, in particular, the transvections belong to SL(V)); the additional requirement that the determinant is 1 means that we must have $\lambda^n = 1$. //

It follows from 2.4.9 that the centre of GL(n, q) is isomorphic to the multiplicative group of the field GF(q), which is cyclic, of order $q - 1$. The centre of SL(n, q) is the subgroup of this group formed by those elements with $\lambda^n = 1$; it is a cyclic group of order $(q - 1, n)$, the highest common factor of $q - 1$ and n.

2.5   Projective spaces and their groups

Our investigations into the geometry of finite vector spaces have
shown us two ways in which that subject just fails to be 'really interesting'.
First, the general linear group $GL(V)$ is not quite 2-transitive on V^*,
since there is no automorphism taking a linearly independent pair of points
into a dependent pair. Secondly, the special linear group $SL(V)$ is not
quite simple, since it may have a nontrivial centre. The classical idea
of a 'projective' geometry shows us how to remove these tiresome imper-
fections: we collapse the vector space by identifying scalar multiples,
and then the central elements of $SL(V)$ have no effect. Of course, projective geometry was originally studied because of its beautiful geometric
structure (every pair of coplanar lines must intersect, and so on), but
this is merely a reflection of the algebraic structure of the underlying
groups.

2.5.1 **Definition.** Let V be a vector space of dimension n over
a field F, and suppose x, $y \in V^* = V - \{0\}$. The statement "$x = \lambda y$
for some $\lambda \in F^*$ " defines an equivalence relation on V^*, and the
equivalence classes are the points of the projective geometry $PG(V)$.

We shall denote the equivalence class of $x \in V^*$ by $[x]$; a
subspace $[U]$ of $PG(V)$ is the image of a subspace U of V under the
map $x \mapsto [x]$. For geometrical reasons it is conventional to say that
if U has dimension k, then $[U]$ has dimension $k - 1$. In particular,
when $V = V(n, q)$ the dimension of $PG(V)$ is $n - 1$, and we write
$PG(V) = PG(n - 1, q)$.

As an example, consider $V = V(3, 2)$. V^* has seven points, and
each of them is the only member of its equivalence class, since F^* has
only one member, 1. Thus $PG(2, 2)$ has seven points. A one-dimension-
al subspace (line) in $PG(2, 2)$ is the image of a 2-dimensional subspace
U of $V(3, 2)$; since U has four points, one of which is the origin, the
line $[U]$ contains just three points. For an explicit representation of
$PG(2, 2)$ we may choose coordinates (x_0, x_1, x_2) for points x in V,
and denote $[x]$ in $PG(2, 2)$ by $[x_0, x_1, x_2]$. Then, for example, the

subspace U whose equation is $x_0 + x_1 + x_2 = 0$ gives rise to a line
[U] containing the points [1, 1, 0], [1, 0, 1], [0, 1, 1]. There are
seven such lines, as depicted in Fig. 2, and we observe that PG(2, 2)
is just the projective plane introduced in Section 2.1. In general,
PG(2, q) is a projective plane satisfying the conditions PP1, PP2, and
PP3.

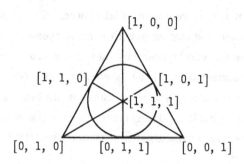

[1, 0, 0]

[1, 1, 0]   [1, 0, 1]

[1, 1, 1]

[0, 1, 0]   [0, 1, 1]   [0, 0, 1]

Fig. 2

We now consider how the action of the general linear group is
affected by passing from V to PG(V). Given g in GL(V), we may
define a permutation $\hat{g}$ of PG(V) by the rule

$$\hat{g}[v] = [g(v)] \quad (v \in V^*).$$

The definition is independent of the chosen representative of [v], since
if [v] = [v'] then v = λv' and g(v) = λg(v'), that is, [gv] = [gv'].
However, the assignment $g \mapsto \hat{g}$ is not a faithful permutation representation of GL(V) on PG(V), since some non-identity automorphisms may
induce the identity on PG(V).

2.5.2 Lemma. <u>Given g in GL(V), the induced permutation $\hat{g}$</u>
<u>is the identity on PG(V) if and only if g is a scalar transformation.</u>

Proof. It is clear from the definition that, if g is scalar, then
$\hat{g}$ is the identity. Conversely, if [gv] = [v] for all v in V*, then
$gv = \lambda_v v$ for some $\lambda_v \in F^*$. The linearity of g implies that λ_v is
independent of v (as in 2.4.9), and so g is a scalar transformation. //

From 2.4.9, we know that the scalar elements of GL(V) and SL(V) comprise the centres of those groups. Thus we may say that the centre of GL(V) (or SL(V)) is the kernel of its permutation representation on PG(V). The quotient of either group by its centre is, in consequence, a group which acts on PG(V).

2.5.3 Definition. The projective general linear group PGL(V), and the projective special linear group PSL(V), are defined as follows:

$$PGL(V) = GL(V)/Z(GL(V)), \quad PSL(V) = SL(V)/Z(SL(V)).$$

When $V = V(n, q)$, it is customary to write the projective groups just defined as PGL(n, q) and PSL(n, q). This convention has the unfortunate consequence that PGL(n, q) acts on PG(n - 1, q) - a somewhat uncomfortable meeting point of algebra and geometry.

2.5.4 Theorem. Both PGL(n, q) and PSL(n, q) act 2-transitively on the points of PG(n - 1, q).

Proof. Given g in GL(n, q), let [g] denote its coset in PGL(n, q), so that the action of the latter group on PG(n - 1, q) is defined by the rule $[g][v] = [g(v)]$ ($v \in V^*(n, q)$). Suppose that [x], [y] and [x'], [y'] are two ordered pairs of distinct points in PG(n - 1, q). The ordered pairs x, y and x', y' are both linearly independent pairs in V(n, q), and so they may be chosen as the initial members of two ordered bases. Hence there is some g in GL(n, q) such that $g(x) = x'$ and $g(y) = y'$; the corresponding [g] takes [x] to [x'] and [y] to [y'], showing that PGL(n, q) acts 2-transitively.

We may obtain an element [h] of PSL(n, q), with the same properties as [g], in the following way. Suppose det $g = \lambda \neq 1$; replace the representative y of [y] by λy, and choose h in GL(n, q) taking x, λy to x', y' (and coinciding with g on the remaining members of the ordered basis). Then det $h = 1$, so that [h] is in PSL(n, q), and it takes [x], [y] to [x'], [y'] (since $[\lambda y] = [y]$). Thus PSL(n, q) also acts 2-transitively. //

An extension of this proof shows that $PGL(n, q)$ and $PSL(n, q)$ are both transitive on triples of non-collinear points ('triangles').

We now set out to prove that the groups $PSL(n, q)$ are simple, provided $n \geq 2$ and $(n, q) \neq (2, 2)$ or $(2, 3)$. The main part of the argument is an extension of the results on $SL(n, q)$ proved in the previous section. The primitivity of the action of $PSL(n, q)$ on $PG(n - 1, q)$, which is a consequence of 2-transitivity, is also important.

2.5.5 Lemma. <u>Let</u> (G, Π) <u>denote the permutation group</u> $PSL(n, q)$ <u>acting on the set</u> $PG(n - 1, q)$ <u>and suppose</u> $\pi \in \Pi$. <u>The stabilizer</u> G_π <u>contains an abelian normal subgroup</u> H_π; <u>the groups</u> H_π ($\pi \in \Pi$) <u>are all conjugate in</u> G <u>and they generate</u> G.

Proof. Suppose that $p \in V(n, q)$ is chosen so that $[p] = \pi$; then the set $T(p)$ of transvections with direction p is independent of the chosen p. Let H_π be the image of $T(p)$ under the epimorphism $SL(n, q) \to PSL(n, q)$. The stated properties of H_π follow from the corresponding properties for $T(p)$ and T proved in 2.4.5 and 2.4.7. //

2.5.6 Lemma. $PSL(n, q)$ <u>coincides with its commutator sub-group, provided</u> $n \geq 2$ <u>and</u> $(n, q) \neq (2, 2)$ <u>or</u> $(2, 3)$.

Proof. This follows from the corresponding result (2.4.8) for $SL(n, q)$, since $PSL(n, q)$ is a quotient of $SL(n, q)$. //

2.5.7 Theorem. <u>The group</u> $PSL(n, q)$ <u>is simple, provided</u> $n \geq 2$ <u>and</u> $(n, q) \neq (2, 2)$ <u>or</u> $(2, 3)$.

Proof. As in 2.5.5, we let (G, Π) denote $PSL(n, q)$ acting on $PG(n - 1, q)$. Suppose that $1 \neq N \trianglelefteq G$; then since G acts 2-transitively, and thus primitively, N must be transitive on Π (by 1.6.6).

Fix $\pi \in \Pi$, let $H = H_\pi$ be as in 2.5.5, and consider the set $NH \subseteq G$. Since N is normal, $NH = HN$ and NH is a subgroup of G. We claim that $NH = G$. To see this, suppose $k \in H_\sigma$, for some $\sigma \in \Pi$, and choose n in N such that $n(\pi) = \sigma$. Then $n^{-1}kn$ is in $H = H_\pi$, and so k is in $NHN = NH$. Since the groups H_σ generate G, we have $NH = G$.

40

We now show that any commutator in G belongs to N. Suppose g_1, $g_2 \in G$; since $G = NH$ we may write $g_1 \in Nh_1$, $g_2 \in Nh_2$ (h_1, $h_2 \in H$). Thus $g_1 g_2 g_1^{-1}$ belongs to

$$Nh_1 Nh_2 h_1^{-1} N = Nh_1 h_2 h_1^{-1} N$$
$$= Nh_2 N$$
$$= Nh_2$$
$$= Ng_2,$$

where we have used the facts that N is normal in G and H is abelian. The calculation shows that $[g_1, g_2]$ belongs to N, and since G coincides with its commutator subgroup, we must have $N = G$. Hence G is simple. //

2.6 More about projective spaces

We shall establish a simple formula for the number of points in $PG(n, q)$. The proof given below is not particularly elegant, but it does provide a useful explicit representation for the points.

2.6.1 Lemma. The number of points in $PG(n, q)$ is $(q^{n+1} - 1)/(q - 1)$.

Proof. The points of $PG(n, q)$ are equivalence classes $[x]$ of points x in $V = V(n + 1, q)$. Suppose that a fixed basis for V is chosen, and that the coordinates of a typical point x are $(x_0, x_1, \ldots, x_n)$. If $x_n \neq 0$, the point $(y_0, y_1, \ldots, 1)$ with $y_i = x_i x_n^{-1}$ ($0 \le i \le n$) is a representative of the point $[x]$ in $PG(n, q)$, and it is the unique representative with last coordinate 1. Since each y_i ($0 \le i \le n - 1$) can be any element of $GF(q)$, there are q^n such points in $PG(n, q)$. Similarly, if $x_{k+1} = \ldots = x_n = 0$, and $x_k \neq 0$, then there is a unique representative of $[x]$ in the form $(y_0, \ldots, y_{k-1}, 1, 0, \ldots, 0)$, and so there are q^{n-k} such points. We have the following classification of the points in $PG(n, q)$:

$$[y_0, y_1, \ldots, y_{n-1}, 1] \qquad q^n \text{ points;}$$

$$[y_0, y_1, \ldots, 1, 0] \qquad q^{n-1} \text{ points;}$$

...   ...

$$[1, 0, \ldots, 0, 0] \qquad 1 \text{ point.}$$

Hence the total number of points is $q^n + q^{n-1} + \ldots + 1 = (q^{n+1}-1)/(q-1)$. //

The representation of the points of the projective line $PG(1, q)$ is especially simple: there are q points $[z, 1]$, $z \in GF(q)$, and one other point $[1, 0]$. Thus there is a one-to-one correspondence β of the set $GF(q) \cup \{\infty\}$ with $PG(1, q)$, defined by

$$\beta(z) = [z, 1] \quad (z \in GF(q)), \quad \beta(\infty) = [1, 0].$$

In practice, we think of the point $[x_0, x_1]$ as corresponding to the 'number' x_0/x_1, considered as an element of $GF(q) \cup \{\infty\}$. In this way, we have a new interpretation of the one-point extension of the affine group over $GF(q)$, given in Section 2. 2. The affine group A consists of the transformations $u \mapsto au + b$ $(a \neq 0)$, and when A is extended by adjoining h (taking u to u^{-1}) we obtain the linear fractional transformations, conventionally written in the form

$$u \mapsto \frac{au + b}{cu + d} \qquad (ad \neq bc) .$$

These transformations operate on $GF(q) \cup \{\infty\}$ with the usual conventions about ∞: $\infty + x = \infty$, $\infty^{-1} = 0$, and so on. The one-to-one correspondence β induces an equivalent permutation group acting on $PG(1, q)$, and the corresponding transformations are represented by 2×2 invertible matrices:

$$\begin{bmatrix} x_0 \\ x_1 \end{bmatrix} \mapsto \begin{bmatrix} a & b \\ c & d \end{bmatrix} \begin{bmatrix} x_0 \\ x_1 \end{bmatrix} \qquad (ad \neq bc) .$$

This shows that the sharply 3-transitive group (P, L) constructed in 2. 2. 3 is equivalent to $PGL(2, q)$ acting on $PG(1, q)$, and we have the following result.

2.6.2 Theorem. PGL(2, q) is sharply 3-transitive on the q + 1 points of PG(1, q). //

We deduce that the order of PGL(2, q) is $(q + 1)q(q - 1)$. In general the orders of PGL(n, q) and PSL(n, q) are easily computed.

2.6.3 Theorem. For all $n \geq 2$ we have

(i) $\quad |PGL(n, q)| = q^{n(n-1)/2} \prod_{i=2}^{n} (q^i - 1);$

(ii) $\quad |PSL(n, q)| = (q - 1, n)^{-1} |PGL(n, q)|,$

where (,) denotes the highest common factor.

Proof. (i) PGL(n, q) is the quotient of GL(n, q) by its centre, which (by 2.4.9) consists of the $q - 1$ scalar transformations $x \mapsto \lambda x$ $(\lambda \neq 0)$. Thus the formula 2.3.4 for $|GL(n, q)|$ yields the result.

(ii) PSL(n, q) is the quotient of SL(n, q) by its centre, which (by 2.4.9) consists of the scalar transformations with $\lambda^n = 1$. Since the non-zero elements of GF(q) form, under multiplication, a cyclic group of order $q - 1$, there are $(q - 1, n)$ such transformations. //

It is instructive to examine the orders of the groups PSL(n, q).

(n, q)	$\|PSL(n, q)\|$	(n, q)	$\|PSL(n, q)\|$
(2, 2)	6	(3, 2)	168
(2, 3)	12	(3, 3)	5616
(2, 4)	60	(3, 4)	20160
(2, 5)	60	. . .	. . .
(2, 7)	168	(4, 2)	20160
(2, 8)	504	. . .	. . .
(2, 9)	360 .		

The two non-simple groups PSL(2, 2) and PSL(2, 3) are isomorphic to S_3 and A_4 respectively. Other possible isomorphisms are suggested by the table, and in fact the following are true:

$PSL(2, 4) \approx PSL(2, 5) \approx A_5;$

$PSL(2, 7) \approx PSL(3, 2);$

$PSL(2, 9) \approx A_6.$

However, the groups of order 20160 require some care, especially when we notice that the simple group A_8 also has this order. The situation is:

$$PSL(3, 4) \not\approx PSL(4, 2); \quad PSL(4, 2) \approx A_8.$$

This is the smallest case where there are two different simple groups with the same order. Remarkably, it can be shown that the instances listed above are the only ones where the orders of symmetric, alternating, and PSL groups coincide.

Finally, we make a few remarks about the geometrical significance of these groups. A permutation of the points of the projective space PG(V) which takes lines to lines is called a collineation. Clearly, if [x] and [y] are collinear and $\hat{g}$ in PGL(V) is induced by g in GL(V), then $\hat{g}[x]$ and $\hat{g}[y]$ are also collinear; thus the elements of GL(V) induce collineations. In fact, it is not necessary to have the strict linearity of g in order to achieve this conclusion.

2.6.4 Definition. Let V be a vector space over a field F. A semilinear automorphism of V is a permutation l of V such that, for some automorphism α of F, we have

$$l(x + y) = l(x) + l(y) \qquad (x, y \in V);$$
$$l(\lambda x) = \alpha(\lambda)l(x) \qquad (x \in V, \lambda \in F).$$

(Of course, if α is the identity, l is linear.)

The group of all semilinear automorphisms of V is denoted by $\Gamma L(V)$; it contains GL(V) as a subgroup. If the underlying field is GF(q), where $q = p^r$, p prime, then there are r field automorphisms and the index $|\Gamma L(V) : GL(V)|$ is equal to r. Each semilinear automorphism l of V induces a permutation $\hat{l}$ of PG(V), in just the same way as for the strict linear automorphisms. Furthermore, the 'shuffling of scalars' involved in the action of l on V is nullified when we pass to PG(V), so that $\hat{l}$ is a collineation of the projective space. The group of these collineations $\hat{l}$ is denoted by $P\Gamma L(V)$.

The 'Fundamental Theorem of Projective Geometry' states that, provided dim $V \geq 3$, the group of all collineations of PG(V) is just

PΓL(V); in other words, every collineation is induced by a semilinear automorphism. This means that the abstract study of projective geometry over finite fields is equivalent to the study of the action of PΓL(V) on PG(V).

The groups PGL(V) and PSL(V) also admit geometrical inter-pretations. We say that a collineation has a <u>centre</u> X if it fixes (setwise) each subspace containing X; it has an <u>axis</u> H (a hyperplane in PG(V)) if it fixes H pointwise. A collineation is an <u>elation</u> if it has a centre X and an axis H such that X ∈ H. The group PGL(V) is just the group generated by all collineations having a centre, while PSL(V) is the group generated by the elations. (The latter result follows from the observation that an elation in PG(V) corresponds to a transvection in V, and the fact that the transvections generate SL(V).)

2.7 The classical simple groups

We have constructed the groups PSL(n, q) and proved that, with a couple of exceptions, they are simple. There are three other types of simple group associated with finite vector spaces: they are known as the symplectic, orthogonal, and unitary groups. In this section we shall sketch the ideas behind the construction of the groups in sufficient detail to enable the reader to work with them, but we shall not attempt to give a thorough treatment. The unifying idea is that of a 'product' or 'form' on a vector space.

2.7.1 Definition. A <u>semibilinear form</u> on a vector space V is a function $s : V \times V \to F$ (where F is the field underlying V), satisfying the conditions:

(i)   for each v in V the function $u \mapsto s(u, v)$ is a linear functional;

(ii)   for each u in V the function $v \mapsto s(u, v)$ is semilinear with respect to some automorphism α of F - that is,
$s(u, x + y) = s(u, x) + s(u, y)$ and $s(u, \lambda x) = \alpha(\lambda) s(u, x)$, $(x, y \in V, \lambda \in F)$.

If the automorphism α in (ii) is the identity, then the form is said to be <u>bilinear</u>. In general, s is <u>nondegenerate</u> if the condition $s(u, v) = 0$ for

45

all v in V implies that u = 0.

Let s : V × V → F be a semibilinear form. For any subspace U
of V, it is clear that

$$\sigma(U) = \{x \in V \mid s(x, u) = 0 \text{ for all } u \in U\}$$

is also a subspace of V. If s is nondegenerate, then σ is a permutation of the set of subspaces, and we say that σ is induced by s. Furthermore, σ <u>reverses inclusion</u>: U ⊆ W implies that σ(W) ⊆ σ(U).

Passing from V to PG(V) by the mapping U ↦ [U] we obtain
an inclusion-reversing permutation of the subspaces of PG(V), also
denoted by σ. Technically, σ is a <u>correlation</u> of PG(V); by invoking
the 'Fundamental Theorem' mentioned in Section 2.6, it can be shown that
any correlation of PG(V) is induced by some semibilinear form on V.

The correlations of special significance for us are those for which
σ^2 is the identity, or equivalently,

$$[W] \subseteq \sigma[U] \Longleftrightarrow [U] \subseteq \sigma[W].$$

Such a correlation is known as a <u>polarity</u>. The condition that a semibilinear form s should induce a polarity is simply that

2.7.2 $s(x, y) = 0 \Longleftrightarrow s(y, x) = 0.$

It turns out that there are only three kinds of semibilinear form s satisfying 2.7.2, and these lead to the three families of simple groups
mentioned above.

In order to describe the forms, it is convenient to pass to an
alternative notation and write ⟨x, y⟩ instead of s(x, y). The types of
semibilinear form satisfying 2.7.2 are:

2.7.3   a <u>symplectic</u> form: ⟨x, x⟩ = 0 for all x in V;

2.7.4   an <u>orthogonal</u> form: ⟨x, y⟩ = ⟨y, x⟩ for all x and y in V;

2.7.5   a <u>unitary</u> form: ⟨x, y⟩ = $\overline{\langle y, x \rangle}$ for all x, y in V, where
         $\lambda \mapsto \overline{\lambda}$ is an automorphism of F with order 2.

2.7.6 Definition. A <u>symplectic</u>, <u>orthogonal</u>, or <u>unitary</u> group on V is a group of all linear automorphisms of V preserving a form of the relevant type: that is,

$$\{g \in GL(V) \,|\, \langle gx, gy \rangle = \langle x, y \rangle \text{ for all } x, y \text{ in } V \}.$$

It is often convenient to use matrix notation in the discussion of particular cases. If we choose a basis $\{e_1, \ldots, e_n\}$ for V, and define a matrix $B = (b_{ij})$ by $b_{ij} = \langle e_i, e_j \rangle$, then B is invertible if the form $\langle \,,\, \rangle$ is nondegenerate. Suppose that the linear automorphism h is represented, with respect to the given basis, by the matrix H. Then h belongs to the group of the form $\langle \,,\, \rangle$ if and only if

$$\langle he_i, he_j \rangle = \langle e_i, e_j \rangle \quad (i, j = 1, \ldots, n).$$

This leads to the matrix equation $B = H^t B H$ in the symplectic and orthogonal cases, and $B = \overline{H}^t B H$ in the unitary case. Taking determinants, we obtain $(\det H)^2 = 1$, and $\det \overline{H} \det H = 1$, in the respective cases.

The number of essentially distinct forms of a given type on a finite vector space $V = V(n, q)$ is small. In the symplectic case, there is just one class: we must have n even, $n = 2m$ say, and a basis for V can be chosen so that

$$\langle x, y \rangle = x_1 y_2 - x_2 y_1 + \ldots + x_{2m-1} y_{2m} - x_{2m} y_{2m-1},$$

where $(x_1, \ldots, x_{2m})$ are the coordinates of x with respect to the chosen basis. In the unitary case, there is likewise just one class: we must have q a square, $q = r^2$ say, and the automorphism $\lambda \mapsto \overline{\lambda}$ of $GF(r^2)$ is given by $\overline{\lambda} = \lambda^r$. A basis for $V(n, r^2)$ can be chosen so that

$$\langle x, y \rangle = x_1 \overline{y}_1 + x_2 \overline{y}_2 + \ldots + x_n \overline{y}_n.$$

The orthogonal case is more complicated, and we shall restrict our discussion to the situation when q is an <u>odd</u> prime power. In this case, the classification of the orthogonal forms $\langle \,,\, \rangle$ is equivalent to the classification of quadratic forms Q, where $Q(x) = \langle x, x \rangle$. When n

is odd there are two classes of forms, but the associated groups

$$\{g \in GL(n, q) \mid Q(gx) = Q(x) \text{ for all } x \in V\}$$

are isomorphic. Thus it is sufficient to consider the typical form

$$Q(x) = x_1 x_2 + x_3 x_4 + \ldots + x_{n-2} x_{n-1} + x_n^2 .$$

When n is even, there are again two classes of quadratic forms, but now the associated groups are not isomorphic. Thus we have to consider the two typical forms

$$Q^+(x) = x_1 x_2 + x_3 x_4 + \ldots + x_{n-1} x_n ,$$
$$Q^-(x) = x_1 x_2 + x_3 x_4 + \ldots + a x_{n-1}^2 + b x_{n-1} x_n + c x_n^2 ,$$

where $a\lambda^2 + b\lambda + c$ is not reducible to linear factors over $GF(q)$.

The classical simple groups are obtained from the groups associated with forms on $V(n, q)$ in much the same way as the groups $PSL(n, q)$ are derived from $GL(n, q)$. We begin with the subgroup consisting of the transformations with determinant 1, and then factor out the centre. In the symplectic case, it can be shown that every transformation preserving a symplectic form must have determinant 1, so that the classical group $PSp(2n, q)$ is obtained by factoring out the centre of the group of the form. The important facts about this family are as follows:

2.7.7   (i) $\displaystyle |PSp(2n, q)| = (q - 1, 2)^{-1} q^{n^2} \prod_{i=1}^{n} (q^{2i} - 1).$

(ii) $PSp(2, q) \approx PSL(2, q).$

(iii) $PSp(2n, q)$ is simple whenever $n \geq 2$, except that $PSp(4, 2)$ is isomorphic to S_6.

In the unitary case, we take the group of those transformations which preserve a unitary form on $V(n, q^2)$ and have determinant 1, and factor out the centre. This yields the family of classical groups $PSU(n, q^2)$, with the following properties:

2.7.8   (i) $\displaystyle |PSU(n, q^2)| = (q + 1, n)^{-1} q^{n(n-1)/2} \prod_{i=2}^{n} [q^i - (-1)^i].$

(ii) $PSU(2, q^2) \approx PSL(2, q).$

(iii) $PSU(n, q^2)$ is simple whenever $n \geq 3$, except that $PSU(3, 2^2)$ is a group of order 72 with a normal subgroup of index 2.

Once again, the orthogonal case is more complicated. Not only do we have to consider the possibility of different forms on the same space, but we also have to use an additional step to obtain the families of simple groups. Furthermore, when q is a power of 2 there are yet more problems. For the details, the reader is referred to one of the standard texts mentioned at the end of the chapter.

There are other infinite families of simple groups which are, in a sense, generalizations of the classical ones. These are the 'simple groups of Lie type'. It would be satisfactory (to those with tidy minds) if it were true that the only finite simple groups are the cyclic groups of prime order, the alternating groups, and the simple groups of Lie type. The fact that it is not true accounts for much of the continuing interest in finite groups. The term 'sporadic' is used to denote a finite simple group which does not belong to one of the known infinite families. We have already encountered two sporadic simple groups: the Mathieu groups M_{11} and M_{12} constructed in Section 1.9. Others will be introduced in the following chapters.

2.8   Project: Near-fields and sharply 2-transitive groups

A near-field is a set E with two binary operations, + and $\otimes$, satisfying the conditions:
    (i) $(E, +)$ is a group, with identity 0;
    (ii) $(E - \{0\}, \otimes)$ is a group;
    (iii) for all a, b, c in E, $(a + b) \otimes c = a \otimes c + b \otimes c$;
    (iv)   for all a in E, $a \otimes 0 = 0$.

2.8.1   Show that the set T of affine transformations

$$u \mapsto u \otimes a + b \quad (a, b \in E, a \neq 0)$$

of a near-field is a group acting sharply 2-transitively on E.

2.8.2   Let $F = GF(q^2)$, where q is a positive power of an odd prime, and let t be a primitive element of F. Define a mapping $b \mapsto \bar{b}$ of $F^* = F - \{0\}$ into Aut F by

$$\bar{b}(a) = \begin{cases} a & \text{if } b = t^s, \ s \text{ even;} \\ a^q & \text{if } b = t^s, \ s \text{ odd.} \end{cases}$$

Show that the new multiplication on F defined by the rule

$$a \otimes b = \bar{b}(a)\, b \quad (b \in F^*), \quad a \otimes 0 = 0,$$

together with the usual addition, makes F into a near-field. (We shall denote this near-field by $N(2, q)$; it is possible to obtain a near-field $N(m, q)$ for certain other pairs (m, q), but we shall not need the general construction.)

2.8.3   Every field $GF(q)$ may be considered as a near-field. Show that the near-field $N(2, 3)$ is <u>not</u> the field $GF(3^2)$.

2.8.4   Let t be a primitive element of $F = GF(3^2)$ satisfying $t^2 + 2t + 2 = 0$. Write down the permutations of F corresponding to the operations $+1$, $+t$, $\otimes t$, $\otimes t^2$ in $N(2, 3)$. Show that these permutations generate the affine group T acting on F considered as the set of elements of the near-field.

2.8.5   Let $B = \{1, 2, \ldots, 9\}$. Find a one-to-one correspondence $\beta : F \to B$ such that the affine group T acting on F is equivalent to the permutation group G_0 acting on B, as in 1.9.8. Show that this group is not equivalent to the affine group A acting on F considered as a field.

2.8.6   Use Theorem 1.5.2 to construct a one-point extension of (T, F), namely, M_{10}.

(In Section 5.4 we shall see that the converse of 2.8.1 holds: any sharply 2-transitive group may be considered as the group of affine transformations of a near-field.)

2.9 Project: Uniqueness of PG(2, 4)

PG(2, 4) has some remarkable properties, and these will be used in the following chapters. In particular, we shall need a uniqueness theorem.

2.9.1 Suppose that $(X, \mathcal{L})$ is a projective plane according to the definition in Section 2.1, and that $|l| = n + 1$ for some line $l \in \mathcal{L}$. Show that every line has $n + 1$ points, and every point belongs to $n + 1$ lines. Deduce that there are $n^2 + n + 1$ points, and $n^2 + n + 1$ lines, in all. (The integer n is the <u>order</u> of the plane.)

2.9.2 A <u>hyperoval</u> in a projective plane of order n is a set of $n + 2$ points, no three of which are collinear. Show that every line meets a given hyperoval in 0 or 2 points.

2.9.3 Let p_1, p_2, p_3, p_4 be four points (no three of them collinear) in a projective plane of order 4, and let l_{ij} be the line determined by p_i and p_j. Show that there are just two points p_5, p_6 not on any l_{ij}, and that $H = \{p_1, p_2, \ldots, p_6\}$ is a hyperoval.

2.9.4 Let X be the set of points of a projective plane of order 4, and suppose that H is a hyperoval. Show that, to each point P of $X - H$, there corresponds a partition of H into three pairs (determined by lines through P), and that every such partition occurs.

2.9.5 Let $(X_i, \mathcal{L}_i)$ be a projective plane of order 4 containing the hyperoval H_i $(i = 1, 2)$. Show that any one-to-one correspondence $H_1 \leftrightarrow H_2$ can be extended to a one-to-one correspondence $X_1 \leftrightarrow X_2$ which induces $\mathcal{L}_1 \leftrightarrow \mathcal{L}_2$. Deduce that any plane of order 4 is isomorphic (in the obvious sense) to PG(2, 4).

2.10 Project: A unitary polarity in PG(2, 9)

The existence of the field automorphism $u \mapsto u^3$ in $GF(3^2)$ allows one to define the unitary form

$$\langle x, y \rangle = x_0 y_0^3 + x_1 y_1^3 + x_2 y_2^3$$

in $V(3, 3^2)$. This induces a polarity in $PG(2, 3^2)$ in the following way. We represent points by symbols $[x_0, x_1, x_2]$ in the usual manner, and the line with equation $l_0 x_0 + l_1 x_1 + l_2 x_2 = 0$ is represented by the symbol $[l_0, l_1, l_2]$. The polarity takes the point $[x_0, x_1, x_2]$ to the line $[x_0^3, x_1^3, x_2^3]$, and the line $[l_0, l_1, l_2]$ to the point $[l_0^3, l_1^3, l_2^3]$.

2.10.1 Show that just 28 of the 91 points in $PG(2, 9)$ lie on their polar lines. These are called <u>isotropic</u> points, and the remaining 63 points are <u>non-isotropic</u>.

2.10.2 Let l be the polar line of an isotropic point. Show that l contains one isotropic point and 9 non-isotropic points.

2.10.3 Let m be the polar line of a non-isotropic point. Show that m contains 4 isotropic points and 6 non-isotropic points.

2.10.4 A <u>self-polar triangle</u> in $PG(2, 9)$ is a set of three points, such that the polar of each point is the line determined by the other two. Show that there are just 63 self-polar triangles.

2.10.5* Find conditions which define the group $PSU(3, 3^2)$ as a group of 3×3 matrices over $GF(3^2)$. Deduce that a matrix belongs to this group if and only if its columns correspond to the three points of a self-polar triangle. Show that the group acts transitively, but not 2-transitively, on the 63 non-isotropic points, and 2-transitively on the 28 isotropic points.

NOTES AND REFERENCES FOR CHAPTER 2

The classical groups were studied by Jordan, and by Dickson [4]. Modern treatments may be found in the books of Artin [1] and Dieudonné [5]. The generalization (mentioned in Section 2.7) to the so-called 'groups of Lie type' is discussed in the book by Carter [2].

Finite projective spaces were discovered by geometers in the nineteenth century, and investigated by Veblen and others in the early 1900s. The books of Dembowski [3] and Hughes and Piper [6] are standard references.

The relationship between near-fields and sharply 2-transitive groups was discovered by Zassenhaus; details are given by Passman [7].

1. Artin, E. Geometric Algebra. (Interscience, 1957.)
2. Carter, R. W. Simple Groups of Lie Type. (Wiley, 1972.)
3. Dembowski, P. Finite Geometries. (Springer-Verlag, 1968.)
4. Dickson, L. E. Linear Groups. (1901, Dover reprint, 1958.)
5. Dieudonné, J. La Géométrie des Groupes Classiques. (Springer-Verlag, 1963.)
6. Hughes, D. R. and Piper, F. C. Projective Planes. (Springer-Verlag, 1973.)
7. Passman, D. S. Permutation Groups. (Benjamin, 1968.)

3 · Designs

'The design of experiments is, however, too large a subject, and of too great importance to the general body of scientific workers, for any incidental treatment to be adequate.'

R. A. Fisher, in the preface to the first edition of The design of experiments, 1935.

3.1   Four fundamental problems

In this chapter we shall be concerned with purely combinatorial structures on a set. There may be geometric overtones, but, as with the abstract notion of a projective plane (Section 2.1), the structures will be defined by numerical properties. In fact, we shall postulate that there is a family $\mathfrak{B}$ of subsets (called blocks) of a set $\mathbf{X}$, satisfying certain regularity conditions. Associated with such a structure there are four fundamental questions:

(i)   Does it exist?

(ii)   Is it unique?

(iii)   Can it be extended to yield a larger structure of the same kind?

(iv)   Does it possess a transitive group of automorphisms?

We have already encountered some aspects of the existence, uniqueness, and extension problems. In particular, we have seen that the 'one-point extension' technique for permutation groups is a powerful tool in the construction of interesting multiply transitive groups. In this chapter, the method will be used to provide extensions of combinatorial structures.

The transitivity problem (iv) is relevant to the basic theme of this book. If a combinatorial structure possesses a group of automorphisms acting transitively (on a set of objects called 'points'), then all points have the same properties and certain numerical regularity conditions will be satisfied. The converse relationship is more problematical. In many cases, numerical regularity does not imply transitivity, but there are some important instances when it does. For example, we recall that the classification of the five regular convex polyhedra is frequently carried out by combinatorial arguments. Numerical conditions lead to just five possible sets of numbers, and each set corresponds to a unique polyhedron. Remarkably, each of the polyhedra is then found to possess a group of automorphisms acting transitively on the vertices. We might say that, in this case, local regularity implies a global symmetry. (In Chapter 5, we shall discuss polyhedra in a more general setting.)

3.2 Designs

A set $Y \subseteq X$, with $|Y| = k$, is often called a k-set, or a k-subset of X, and the set of all k-subsets of X is denoted by $X^{(k)}$. Consider the following collection of 5-subsets of the 11-set $X = \{0, 1, 2, \ldots, 9, T\}$.

0 2 3 4 8	1 4 6 7 8	0 1 5 8 T
1 3 4 5 9	2 5 7 8 9	0 1 2 6 9
2 4 5 6 T	3 6 8 9 T	1 2 3 7 T.
0 3 5 6 7	0 4 7 9 T	

It can be verified that each 2-subset of X occurs exactly twice as a subset of one of the given 5-subsets. This is an example of a 'design'.

3.2.1 Definition. A design with parameters $t - (v, k, \lambda)$ is a pair $(X, \mathcal{B})$, such that

(i) X is a v-set;

(ii) $\mathcal{B}$ is a collection of k-subsets of X;

(iii) each t-subset of X is contained in exactly λ members of $\mathcal{B}$.
The elements of X are called points, and the elements of $\mathcal{B}$, blocks. We shall assume that all the parameters are positive integers, and that $v > k \geq t$ (to avoid trivial cases). Also, the members of $\mathcal{B}$ must be distinct: repeated blocks are not allowed.

We often speak of a $t - (v, k, \lambda)$ design, or a t-design, with the obvious meaning. In practice, we are interested in t-designs with $t \geq 2$, although it is convenient to retain the terminology for the case $t = 1$.

The example above is a $2 - (11, 5, 2)$ design, and the seven-point plane of Section 2.1 is a $2 - (7, 3, 1)$ design. Another simple example is the <u>complete</u> design, where $\mathcal{B} = X^{(k)}$; this is a $k - (v, k, 1)$ design.

3.2.2 Theorem. <u>A</u> t-<u>design</u> $(X, \mathcal{B})$ <u>is also an</u> s-<u>design, for</u> <u>each value of</u> s <u>in the range</u> $1 \le s \le t$. <u>If the given design has para-</u> <u>meters</u> $t - (v, k, \lambda)$, <u>then its parameters as an</u> s-<u>design are</u> $s - (v, k, \lambda_s)$, <u>where</u>

$$\lambda_s = \lambda \cdot \frac{(v - s)(v - s - 1) \dots (v - t + 1)}{(k - s)(k - s - 1) \dots (k - t + 1)} .$$

Proof. We proceed by induction on $t - s$; clearly, the result is true when $t - s = 0$.

Suppose that the result holds when $s = i + 1$ $(1 \le i \le t - 1)$, so that each $(i + 1)$-subset of X occurs as a subset of exactly λ_{i+1} blocks. Let I be any i-subset of X and consider the pairs $(x, \mathcal{B}) \in X \times \mathcal{B}$ satisfying the conditions

$$x \in X - I, \quad I \cup \{x\} \subseteq \mathcal{B}.$$

By the counting principle 1.2.2, we have

$$(v - i)\lambda_{i+1} = (k - i)\lambda_i(I),$$

where $\lambda_i(I)$ is the number of blocks containing I, and the equation shows that this is a constant λ_i, independent of I. Thus we have an i-design, and the expression for λ_i follows by repeated application of the preceding equation. The induction step is complete, and the result holds for all s in the range $1 \le s \le t$. //

The symbol λ_1, which denotes the number of blocks containing any given point, is conventionally replaced by r. Furthermore, the argument in the theorem remains true when $i = 0$; in fact λ_0, the number of blocks containing the empty set, is simply $|\mathcal{B}|$, and is usually denoted by b. Thus we have the general equation

56

3.2.3 $(v - i)\lambda_{i+1} = (k - i)\lambda_i$ $(0 \le i \le t - 1)$,

and the important case $i = 0$ may be written

3.2.4 $vr = bk$.

If the parameters $t - (v, k, \lambda)$ are given, 3.2.3 enables us to calculate $\lambda_t = \lambda$, λ_{t-1}, ..., $\lambda_1 = r$, and $\lambda_0 = b$ in turn. The fact that these must be integers gives the so-called 'divisibility conditions' for the existence of a design.

    3.2.5 Theorem.   <u>In order that a design with parameters</u> $t - (v, k, \lambda)$ <u>may exist, it is necessary that</u>

$$(k - i)(k - i - 1) \ldots (k - t + 1) \text{ divides } \lambda(v - i)(v - i - 1) \ldots (v - t + 1),$$

<u>for each</u> i <u>in the range</u> $0 \le i \le t - 1$. //

    For example, there cannot be a $3 - (11, 4, 1)$ design, since the condition with $i = 2$ is not satisfied. However, there could be a $5 - (28, 7, 1)$ design, since we find that $\lambda_4 = 8$, $\lambda_3 = 50$, $\lambda_2 = 260$, $\lambda_1 = r = 1170$, $\lambda_0 = b = 4680$. In Section 3.8 we shall see that, in this case, there is indeed a design.

    We remark that, if the parameters t, v, k, and b are given, then $\lambda_0 = b$, $\lambda_1 = r$, λ_2, ..., $\lambda_t = \lambda$ may be calculated by reversing the preceding arguments.

    At this point, it may be helpful to mention some of the many special terms occurring in the literature on designs. A <u>Steiner system</u> is simply a $t - (v, k, \lambda)$ design with $t \ge 2$ and $\lambda = 1$; the notation $S(t, k, v)$ is sometimes used. An $S(2, 3, v)$ is a <u>Steiner triple system</u>. In statistics, the term <u>BIBD</u> (balanced incomplete block design) is often found; it denotes a 2-design which is not complete, that is, $b \ne \binom{v}{k}$. (We remark that repeated blocks are frequently allowed in statistical applications.) It can be shown (see 2.9.1) that the definition of a finite projective plane in Section 2.1 implies that $(X, \mathcal{L})$ is a $2 - (n^2 + n + 1, n + 1, 1)$ design for some integer n. The converse is also true, and consequently the term <u>projective plane</u> of <u>order</u> n is often used for a design with these parameters. PG(2, q) is a projective plane of order q, but there are other projective planes; for example, there are at least three different projective planes of order 9, in addition to PG(2, 9).

In order to obtain stronger non-existence results for designs, we introduce the following definition.

3.2.6 **Definition.** Let $(X, \mathcal{B})$ be a design with

$$X = \{x_1, x_2, \ldots, x_v\}, \quad \mathcal{B} = \{\beta_1, \beta_2, \ldots, \beta_b\}.$$

The <u>incidence matrix</u> of $(X, \mathcal{B})$ is the $v \times b$ matrix $A = (a_{ij})$ defined by

$$a_{ij} = \begin{cases} 1 & \text{if } x_i \in \beta_j, \\ 0 & \text{if } x_i \notin \beta_j. \end{cases}$$

Even for a 2-design, the algebraic properties of this simple matrix lead to nontrivial combinatorial consequences. We shall suppose for the rest of this section that $(X, \mathcal{B})$ is a $2 - (v, k, \lambda)$ design, with $\lambda_1 = r$ and $\lambda_0 = b$, so that the five parameters are related by the equations $vr = bk$ and $(v - 1)\lambda = (k - 1)r$.

3.2.7 **Lemma.** <u>Let</u> I <u>denote the</u> $v \times v$ <u>identity matrix and</u> J <u>the</u> $v \times v$ <u>matrix each entry of which is</u> 1. <u>The incidence matrix</u> A <u>of</u> <u>a</u> $2 - (v, k, \lambda)$ <u>design satisfies</u>

$$AA^t = (r - \lambda)I + \lambda J.$$

Proof. $(AA^t)_{ij} = \sum a_{i\ell} a_{j\ell}$, and this is just the number of blocks β_ℓ which contain both x_i and x_j. If $i = j$, this number is r; if $i \neq j$ it is λ. $/\!/$

3.2.8 **Lemma.** <u>The incidence matrix of a 2-design satisfies</u>

$$\det(AA^t) = rk(r - \lambda)^{v-1}.$$

Proof. We have just proved that $AA^t = (r - \lambda)I + \lambda J$. Subtract the first column from each other column, and then add each row to the first row. The result is a triangular matrix with diagonal entries

58

$r + (v-1)\lambda$, $r - \lambda$, ..., $r - \lambda$. Since $(v-1)\lambda = (k-1)r$ and the determinant is the product of the diagonal entries, we have the result. $/\!/$

3.2.9 Theorem. <u>If a 2-design on</u> v <u>points has</u> b <u>blocks, then</u> $b \geq v$.

Proof. The determinant of AA^t is zero only if $r = \lambda$. But $r = \lambda$ implies $k = v$, which is forbidden by our definition, so that AA^t is invertible, and its rank is v. The rank of the product of two matrices is not greater than the rank of either factor, and the rank of A is not greater than the number b of its columns. Hence

$$b \geq \mathrm{rank}(A) \geq \mathrm{rank}(AA^t) = v. \quad /\!/$$

We note that this result means that $b \geq v$ for every t-design with $t \geq 2$, since each such design is also a 2-design.

3.3 Symmetric designs

3.3.1 Definition. A $t - (v, k, \lambda)$ design is <u>symmetric</u> if it is incomplete and $t \geq 2$, $b = v$.

The crucial condition is that $b = v$. For various reasons, the term 'symmetric' is not particularly apt, but it has become conventional and so we shall use it.

3.3.2 Theorem. <u>If a</u> $t - (v, k, \lambda)$ <u>design is symmetric, then</u> $t = 2$.

Proof. Let $(X, \mathcal{B})$ be a symmetric $t - (v, k, \lambda)$ design, and suppose that $t \geq 3$. Then by 3.2.2, the design is also a $3 - (v, k, \lambda')$ design, for some λ'. If we remove one point x from the blocks which contain it, and delete all the blocks not containing x, the result is a $2 - (v-1, k-1, \lambda')$ design. This has r blocks (where r is the number of occurrences of x in the original 3-design), and by 3.2.9 we must have $r \geq v - 1$. But in the 3-design $b = v$, and so $r = k$ (3.2.4). Thus we must have $k \geq v - 1$, and the only possibility is that $(X, \mathcal{B})$ is the complete $(v - 1) - (v, v - 1, 1)$ design, which is not a symmetric design by the terms of the definition. //

We remark that for a symmetric $2 - (v, k, \lambda)$ design the five basic parameters b, v, r, k, λ are related by the equations

3.3.3 $b = v, \; r = k, \; k(k - 1) = \lambda(v - 1)$.

3.3.4 **Lemma.** <u>If</u> A <u>is the incidence matrix of a symmetric</u> $2 - (v, k, \lambda)$ <u>design, we have not only</u> $AA^t = (k - \lambda)I + \lambda J$ <u>(as in 3.2.7),</u> <u>but also</u>

$$A^t A = (k - \lambda)I + \lambda J.$$

Proof.   Since each point occurs in $r = k$ blocks, we have $AJ = kJ$; and since each block contains k points, $JA = kJ$. Now $\det(AA^t) \neq 0$ (as in 3.2.9), and in this case A is a square matrix so that A^{-1} exists. Multiplying 3.2.7 on the left by A^{-1}, we get

$$A^t = A^{-1}AA^t = (k - \lambda)A^{-1} + \lambda A^{-1}J;$$

but $A^{-1}J = A^{-1}(k^{-1}AJ) = k^{-1}J$, and so

$$A^t A = (k - \lambda)A^{-1}A + \lambda k^{-1}JA$$
$$= (k - \lambda)I + \lambda J. \; //$$

3.3.5 **Theorem.** <u>If</u> $(X, \mathcal{B})$ <u>is a symmetric</u> $2 - (v, k, \lambda)$ <u>design,</u> <u>then any two distinct blocks intersect in exactly</u> λ <u>points.</u>

Proof.   We have $(A^t A)_{ij} = \sum a_{\ell i} a_{\ell j}$, which is equal to the number of points x_ℓ belonging to both β_i and β_j. But, if $i \neq j$, the preceding lemma shows that $(A^t A)_{ij} = \lambda$. Hence the result. //

We have shown that in a symmetric design, not only is it true (by definition) that each pair of points belongs to λ blocks, but also it is true that each pair of blocks intersects in just λ points. This 'duality' may be formulated in the following way.

3.3.6 Definition. Suppose that $(X, \mathcal{B})$ is a design, and $x \in X$. The $\underline{\text{star}}$ of x is the set of blocks containing X, and it is denoted by $\text{st}(x)$. The $\underline{\text{dual}}$ of the design $(X, \mathcal{B})$ is $(X^*, \mathcal{B}^*)$, where $X^* = \mathcal{B}$ and

$$\mathcal{B}^* = \{\text{st}(x) | x \in X\}.$$

By the definition of a design, it follows that if the parameters of $(X, \mathcal{B})$ as a 1-design are $1 - (v, k, r)$, then $(X^*, \mathcal{B}^*)$ is a $1 - (b, r, k)$ design. Thus we may speak of the $\underline{\text{dual design.}}$ In general, the dual of a t-design with $t \geq 2$ will be an s-design only for $s = 1$; however Theorem 3.3.5 shows that the dual of a symmetric $2 - (v, k, \lambda)$ design is also a $2 - (v, k, \lambda)$ design. For example, the seven-point plane (Section 2.1) is a $2 - (7, 3, 1)$ design on the set C of integers modulo 7, with blocks $\beta_i = \{i, i + 1, i + 3\}$ $(i \in X)$: the dual is a $2 - (7, 3, 1)$ design on the set $\mathcal{B} = \{\beta_i\}$, with blocks $\beta_j^* = \text{st}(j) = \{\beta_j, \beta_{j+4}, \beta_{j+6}\}$ $(j \in X)$. In this particular case the dual is isomorphic to the original design, in the sense that the mapping $i \mapsto \beta_{-i}$ takes the blocks of the first design to the blocks of the second.

One should note that not all symmetric designs are self-dual. In the case of projective planes of order n (or $2 - (n^2 + n + 1, n + 1, 1)$ designs), the standard plane $PG(2, n)$ is self-dual when it exists - that is when n is a prime power. But there is, for example, a non-standard plane of order 9 which is not self-dual.

We now turn to the existence problem for symmetric designs and, in particular, for projective planes. The planes $PG(2, n)$ provide examples of projective planes of order n for $n = 2, 3, 4, 5, 7, 8, 9, \ldots$, and in fact these are the unique planes for $n \leq 8$. When $n = 9$ there are at least four different planes: $PG(2, 9)$ and one other self-dual plane, and two planes dual to each other. The question of a possible plane of order 6 will be settled by the following theorems, but the question of a plane of order 10 is still unresolved (May 1978).

3.3.7 Theorem. $\underline{\text{If}}$ v $\underline{\text{is an even integer and a symmetric}}$ $2 - (v, k, \lambda)$ $\underline{\text{design exists, then}}$ $k - \lambda$ $\underline{\text{must be a perfect square.}}$

Proof. Let A be the incidence matrix of the design. Then, using 3. 2. 8 and 3. 3. 3, we have

$$(\det A)^2 = \det (AA^t)$$
$$= rk(r - \lambda)^{v-1}$$
$$= k^2(k - \lambda)^{v-1}.$$

Thus $(k - \lambda)^{v-1}$ is a perfect square, and if $v - 1$ is odd then $k - \lambda$ itself must be a perfect square. $/\!/$

This result shows that the simple basic identities 3. 3. 3 are not sufficient to ensure the existence of a $2 - (v, k, \lambda)$ design. For example, the numbers $v = 46$, $k = 10$, $\lambda = 2$ satisfy the basic identities, but v is even and $k - \lambda$ is not a square so that there is (by 3. 3. 7) no design.

The corresponding result for the case when v is odd requires some preliminary discussion. We shall have to assume, from elementary number theory, the fact that every prime number can be written as the sum of (at most) four squares. In order to extend this result to composite numbers, let h_1, h_2, h_3, h_4 be integers (not all zero), and consider the following matrix:

$$H = \begin{bmatrix} h_1 & h_2 & h_3 & h_4 \\ h_2 & -h_1 & h_4 & -h_3 \\ h_3 & -h_4 & -h_1 & h_2 \\ h_4 & h_3 & -h_2 & -h_1 \end{bmatrix}.$$

We have $HH^t = (h_1^2 + h_2^2 + h_3^2 + h_4^2)I$, so that if $x = [x_1, \ldots, x_4]$ and $y = xH = [y_1, \ldots, y_4]$ then

$$y_1^2 + y_2^2 + y_3^2 + y_4^2 = yy^t = x(HH^t)x^t$$
$$= (h_1^2 + h_2^2 + h_3^2 + h_4^2)(x_1^2 + x_2^2 + x_3^2 + x_4^2).$$

Thus, a product of two sums of four squares is also a sum of four squares, so that any number (being a product of primes) can be expressed as a sum of four squares. Furthermore, the argument shows that each of the expressions $x_1, \ldots, x_4$ may be written as a rational linear combination of $y_1, \ldots, y_4$; this is because H is invertible, $x = H^{-1}y$, and the entries

62

of H^{-1} are rational numbers.

3.3.8 Theorem. <u>If v is an odd integer and a symmetric</u>
<u>$2 - (v, k, \lambda)$ design exists, then the equation</u>

$$x^2 = (k - \lambda)y^2 + (-1)^{\frac{1}{2}(v-1)}\lambda z^2$$

<u>must have a solution in integers,</u> x, y, z, <u>not all zero.</u>

Proof. Suppose that a design exists; then we have the identity
$AA^t = (k - \lambda)I + \lambda J$ (3.2.7 with $r = k$) for the incidence matrix. The
idea of the proof is to interpret this as an identity in quadratic forms over
the rational field.

If x is the row vector $[x_1, \ldots, x_v]$, then the identity for AA^t
gives

$$xAA^t x^t = (k - \lambda)(x_1^2 + \ldots + x_v^2) + \lambda(x_1 + \ldots + x_v)^2.$$

Putting $f = xA$, we have $ff^t = xAA^t x^t$ and

(*) $f_1^2 + \ldots + f_v^2 = (k - \lambda)(x_1^2 + \ldots + x_v^2) + \lambda(x_1 + \ldots + x_v)^2.$

The equation (*) is an identity in $x_1, \ldots, x_v$. Each of the f's is a
rational linear combination of the x's, since $f = xA$.

Suppose first that $v \equiv 1 \pmod 4$. We express the integer $k - \lambda$
as the sum of four squares, and bracket the terms of $x_1^2 + \ldots + x_{v-1}^2$ in
fours. Each product of sums of four squares is itself a sum of four
squares, and so (*) yields

(**) $f_1^2 + \ldots + f_v^2 = y_1^2 + \ldots + y_{v-1}^2 + (k - \lambda)y_v^2 + \lambda z^2$,

where $y_v = x_v$, $z = x_1 + \ldots + x_v$, and the y's are related to the x's
by an invertible linear transformation with rational coefficients. Since
the f's are rational linear combinations of the x's, it follows that the
f's (and z) are rational linear combinations of the y's.

Suppose that $f_1 = b_{11}y_1 + \ldots + b_{1v}y_v$. We can define y_1 as a
rational linear combination of $y_2, \ldots, y_v$, in such a way that $f_1^2 = y_1^2$:
if $b_{11} \neq 1$ we set $y_1 = (1 - b_{11})^{-1}(b_{12}y_2 + \ldots + b_{1v}y_v)$, while if

63

$b_{11} = 1$ we set $y_1 = (-1 - b_{11})^{-1}(b_{12}y_2 + \ldots + b_{1v}y_v)$. Now we know that f_2 is a rational linear combination of the y's, and, using the relevant expression for y_1 found above, we can express f_2 as a rational linear combination of $y_2, \ldots, y_v$. As before, we fix y_2 as a function of $y_3, \ldots, y_v$ in such a way that $f_2^2 = y_2^2$. Continuing thus, we eventually obtain $y_1, \ldots, y_{v-1}$ and $f_1, \ldots, f_v$ as rational functions of y_v, satisfying $f_i^2 = y_i^2$ $(1 \le i \le v-1)$.

Choose any non-zero rational value for y_v. All the f's, the remaining y's, and z, are determined as above, and substituting these values in (**) we obtain

$$f_v^2 = (k - \lambda)y_v^2 + \lambda z^2.$$

Multiplying by a suitable constant we have an integral solution for the equation, and the theorem is proved in the case $v \equiv 1 \pmod 4$.

If $v \equiv 3 \pmod 4$ we add an extra term $(k - \lambda)x_{v+1}^2$ to the basic identity (*), and repeat the argument given above. //

In order to use Theorem 3.3.8 to show that symmetric designs with certain parameters cannot exist, we must show that the corresponding equations have no integral solutions. We could do this by quoting some general results from number theory, but for our purposes an illustrative example will suffice.

3.3.9 Theorem. <u>There is no projective plane of order 6.</u>

Proof. A projective plane of order 6 is a symmetric 2 - (43, 7, 1) design. In view of 3.3.8, we have to show that the equation

$$x^2 = 6y^2 - z^2$$

has no solution in integers x, y, z, not all zero.

If a solution exists, we may, without loss of generality, suppose that x, y, and z have no common factor. Since $x^2 + z^2 = 6y^2$, $x^2 + z^2$ is divisible by 3. Let r $(= 0, 1, 2)$ be the residue of z modulo 3; then $x^2 + r^2$ is also divisible by 3. If $r = 1$ or $r = 2$, we must have $x^2 + 1$ divisible by 3, which is impossible for any integer x. Thus

$r = 0$. In other words, 3 divides z^2, and so 3 divides z also. Similarly, 3 divides x, so that $x^2 + z^2$ is divisible by 9. But this means that y^2 and y must be divisible by 3, which contradicts the assumption that x, y, z, have no common factor.

Hence the equation has no solution and there is no projective plane of order 6. $/\!/$

Similar methods may be used to rule out the possibility of some other projective planes, for example those of order 14 and 21. As we have already remarked, the question of the plane of order 10 is still open: in fact, all known planes have prime power order!

3.4   Automorphisms of designs

3.4.1 **Definition.** An _automorphism_ of a design $(X, \mathcal{B})$ is a permutation π of X such that $\beta \in \mathcal{B}$ implies $\pi(\beta) \in \mathcal{B}$.

Clearly, the automorphisms of $(X, \mathcal{B})$ form a group which acts on X. Since an automorphism takes blocks to blocks, the group also has a permutation representation on the set $\mathcal{B}$.

3.4.2 **Lemma.** Let $(X, \mathcal{B})$ be a t-design with $t \geq 2$. Then the group of automorphisms of the design acts faithfully on $\mathcal{B}$.

Proof. Suppose that g is an automorphism of $(X, \mathcal{B})$ such that $g(\beta) = \beta$ for every block β; we have to show that g is the identity permutation on X. For any x in X, g takes the set $st(x)$ of blocks containing x to the set $st(gx)$, and since g fixes each block setwise, we have $st(gx) = st(x)$. Let r and λ have their usual meanings for $(X, \mathcal{B})$, considered as a 2-design. Suppose $gx \neq x$; the condition $st(gx) = st(x)$ says that the number (r) of blocks containing x is equal to the number (λ) of blocks containing gx and x. But $r = \lambda$ implies $v = k$, which is not allowed, and so $gx = x$ for all $x \in X$. $/\!/$

The existence of multiply transitive groups may be used in order to construct designs, as in the following theorem.

3.4.3 Theorem. Let (G, X) be a t-transitive permutation group $(t \geq 2)$, and suppose β is a subset of X, with $|\beta| = k$, $|X| = v$, and $1 < k < v-1$. Then the set

$$\mathcal{B} = \{g(\beta) \, | \, g \in G\}$$

is the set of blocks of a t-design $(X, \mathcal{B})$, and G is a group of automorphisms acting transitively on $\mathcal{B}$.

Proof. If S and T are any two t-subsets of X, then there is some h in G such that $h(S) = T$. Whenever S belongs to a block $g(\beta)$ in $\mathcal{B}$, T belongs to the block $hg(\beta)$. Thus S and T occur equally often as subsets of a block. //

The parameters of the design $(X, \mathcal{B})$ constructed in 3.4.3 are $t - (v, k, \lambda)$, where λ may be found as follows. The number $b = |\mathcal{B}|$ is simply the length of the orbit of β in the action of G on the set $X^{(k)}$ of all k-subsets of X. Thus $b = |G:G_{(\beta)}|$, where $G_{(\beta)}$ is the setwise stabilizer of β. From this, the parameters $r = \lambda_1, \lambda_2, \ldots, \lambda_t = \lambda$ may be found, using 3.2.3 and 3.2.4.

We remark that $\mathcal{B}$ depends not only on G and k, but also on the specific k-subset β which is chosen. For some choices of β, we may obtain a complete design - that is, $b = \binom{v}{k}$.

As an example of 3.4.3 consider the sharply 2-transitive group A of affine transformations, acting on the field $F = GF(9)$. Let t be a primitive element satisfying $t^2 = t + 1$ (see Section 2.2). If we choose $\beta = \{0, t, t^2\}$, then $|A_{(\beta)}| = 1$, and there are $|A:A_{(\beta)}| = 72$ blocks. However, if we choose $\beta = \{0, t, t^5\}$, we find that the stabilizer $A_{(\beta)}$ is generated by $u \mapsto u + t$ and $u \mapsto 2u$, so that $|A_{(\beta)}| = 6$ and the number of blocks generated by the action of G on β is $b = |A:A_{(\beta)}| = 72/6 = 12$. In this case $r = 4$ and $\lambda = 1$, so we have constructed a $2 - (9, 3, 1)$ design, or a Steiner triple system on 9 points.

3.5 Extensions of designs

The following table gives the 14 blocks of a $3 - (8, 4, 1)$ design on the set $X = \{0, 1, \ldots, 7\}$:

```
0 1 2 4          3 5 6 7
0 2 3 5          1 4 6 7
0 3 4 6          1 2 5 7
0 4 5 7          1 2 3 6
0 1 5 6          2 3 4 7
0 2 6 7          1 3 4 5
0 1 3 7          2 4 5 6 .
```

It will be seen that if the blocks not containing 0 are removed, and if 0 is deleted from those which remain, the result is the familiar 2 - (7, 3, 1) design. In fact, this is a particular case of a general construction. If $D = (X, \mathcal{B})$ is a t - (v, k, λ) design, and $x \in X$, then the family of sets

$$\mathcal{B}_x = \{\beta - \{x\} \,|\, x \in \beta, \ \beta \in \mathcal{B}\}$$

gives a block design on $X - \{x\}$ with parameters (t-1) - (v-1, k-1, λ). The design $D_x = (X - \{x\}, \mathcal{B}_x)$ is called a <u>contraction</u> of D. Since the number b_x of blocks of D_x is equal to the number r of blocks of D which contain x, we have

$$3.5.1 \quad b_x = bk/v.$$

We shall be concerned with the possibility of reversing the contraction process, so that designs with larger values of t may be constructed. There are remarkably few cases where the construction is possible, but the rarity of success makes it even more interesting.

3.5.2 Definition. The design $D^+ = (X^+, \mathcal{B}^+)$ is an <u>extension</u> of $D = (X, \mathcal{B})$ if $X^+ = X \cup \{z\}$ for some point $z \notin X$, and the contraction $(D^+)_z$ is just D. D is said to be <u>extendable</u> if it has some extension D^+.

The example given above shows that the 2 - (7, 3, 1) design is extendable. (In fact, any 2 - (4λ+3, 2λ+1, λ) design is extendable - see 3.7.) A simple, but useful, criterion for extendability is given in the following lemma.

3.5.3 Lemma. A necessary condition for a $t - (v, k, \lambda)$ design with b blocks to be extendable is that $k + 1$ divides $b(v + 1)$.

Proof. Suppose that there is an extension; it is a $(t + 1) - (v + 1, k + 1, \lambda)$ design with b^+ blocks, where (by 3.5.1)

$$b = b^+(k + 1)/(v + 1).$$

Thus, in order that b^+ should be an integer, $k + 1$ must divide $b(v+1)$. //

3.5.4 Theorem. Let q be a prime power. A necessary condition that a design with parameters $2 - (q^2 + q + 1, q + 1, 1)$ — in particular $PG(2, q)$ — should be extendable is that $q = 2$ or 4.

Proof. By the lemma, a necessary condition for extendability is that $q + 2$ should divide $(q^2 + q + 1)(q^2 + q + 2)$. Now

$$(q^2 + q + 1)(q^2 + q + 2)/(q + 2) = (q^3 + 4q - 5) + 12/(q + 2),$$

so that $q + 2$ must divide 12, and since q is a prime power this means that $q = 2$ or 4. //

In fact, both $PG(2, 2)$ and $PG(2, 4)$ are extendable, and there is a unique extension in each case (see 3.7.5 and 3.9 respectively). $PG(2, 2)$ is a $2 - (7, 3, 1)$ design, and its extension is displayed at the beginning of this section. $PG(2, 4)$ is a $2 - (21, 5, 1)$ design, and the extension is a $3 - (22, 6, 1)$ design. Remarkably, this can itself be extended twice more, so that we obtain $4 - (23, 7, 1)$ and $5 - (24, 8, 1)$ designs. We shall construct these extensions by simultaneously extending the automorphism group of the design, using the one-point extension technique of 1.5.2.

The procedure is outlined in the flow diagram opposite. We begin with a given $t - (v, k, \lambda)$ design $(X, \mathcal{B})$, for which the automorphism group G acts t-transitively on X and transitively on $\mathcal{B}$. Then we suppose that (G, X) admits a one-point extension (G^+, X^+), where $X^+ = X \cup \{z\}$. Finally, we apply Theorem 3.4.3, taking the basic block β to be $\beta_0 \cup \{z\}$, for some β_0 in $\mathcal{B}$. By this means we construct a $(t + 1) - (v + 1, k + 1, \lambda^+)$ design $(X^+, \mathcal{B}^+)$; however, it is not necessarily an extension of $(X, \mathcal{B})$ since, for example, λ^+ may not be the same as λ.

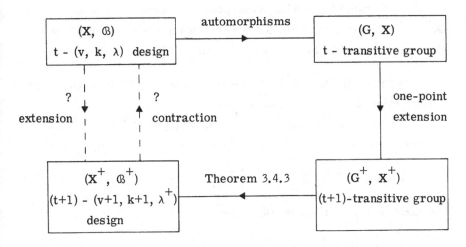

As an example of the procedure, let us take $(X, \mathcal{B})$ to be the 2 - (9, 3, 1) design constructed in the preceding section. We know that the affine group acts 2-transitively on the point-set, $GF(9)$, of the design, and that it is transitive on the blocks. Also, we recall (from Sections 2.2 and 2.5) the fact that the group has a one-point extension: it consists of the linear fractional transformations acting sharply 3-transitively on the projective line $PG(1, 9) = GF(9) \cup \{\infty\}$. Thus we may invoke Theorem 3.4.3 to construct a 3 - (10, 4, λ^+) design. We take the basic block β to be $\{0, t, t^5, \infty\}$, since we know that $\beta_0 = \{0, t, t^5\}$ is a block of the original design. The setwise stabilizer $G_{(\beta)}^+$ contains the linear fractional transformation $u \mapsto (tu + t^2)/u$ and the setwise stabilizer of β_0 (which has order 6); thus $|G_{(\beta)}^+| = 24$. We deduce that

$$|\mathcal{B}^+| = |G^+ : G_{(\beta)}^+| = 720/24 = 30,$$

and $r = 12$, $\lambda_2 = 4$, $\lambda_3 = \lambda^+ = 1$. Since the new λ^+ is the same as the original λ, the new design could be an extension of the old one. In fact, the condition $\lambda^+ = \lambda$ is always sufficient for this to be so.

69

3.5.5 Theorem. <u>Let $(X, \mathcal{B})$ be a t - (v, k, λ) design whose automorphism group G is transitive on blocks, and suppose that (G, X) is t-transitive and has a one-point extension. The design $(X^+, \mathcal{B}^+)$ constructed by the technique shown in the flow diagram is an extension of $(X, \mathcal{B})$ if and only if $\lambda^+ = \lambda$.</u>

Proof. If $(X^+, \mathcal{B}^+)$ is an extension of $(X, \mathcal{B})$, then $(X, \mathcal{B})$ is a contraction of $(X^+, \mathcal{B}^+)$ and $\lambda^+ = \lambda$.

Conversely, suppose that $\lambda^+ = \lambda$. Applying 3.2.2 to $(X^+, \mathcal{B}^+)$ we find

$$r^+ = \lambda^+ \cdot \frac{v(v-1) \ldots (v-t+1)}{k(k-1) \ldots (k-t+1)} .$$

Similarly, applying 3.2.2 to $(X, \mathcal{B})$ we have

$$r = \lambda \cdot \frac{(v-1) \ldots (v-t+1)}{(k-1) \ldots (k-t+1)} .$$

Thus $(r^+/r) = (\lambda^+/\lambda) \cdot (v/k)$, and if $\lambda = \lambda^+$ we have $r^+ = vr/k = bk/k = b$. Since z occurs r^+ times in the blocks $\mathcal{B}^+$, it follows that the contraction with respect to z of $(X^+, \mathcal{B}^+)$ must have $r^+ = b$ blocks. We must show that these are precisely the blocks in $\mathcal{B}$. Suppose the basic block used in the construction of $(X^+, \mathcal{B}^+)$ is $\beta = \beta_0 \cup \{z\}$, for some $\beta_0 \in \mathcal{B}$, and suppose γ is any member of $\mathcal{B}$. Since G is transitive on $\mathcal{B}$, we have $\gamma = g(\beta_0)$ for some g in G. Thus $\gamma \cup \{z\} = g(\beta)$, and $\gamma \cup \{z\}$ is in $\mathcal{B}^+$, since $g \in G^+$. Contracting $\mathcal{B}^+$ with respect to z gives the block γ. In other words $(\mathcal{B}^+)_z = \mathcal{B}$, as required. $/\!/$

3.6 Mathieu groups and associated designs

In this section we shall apply the extension procedure introduced above, beginning with the 2 - $(21, 5, 1)$ design $PG(2, 4)$ and the group $PSL(3, 4)$. Three successive one-point extensions of $PSL(3, 4)$ will be constructed, using the method of Theorem 1.5.2.

We shall represent points of $PG(2, 4)$ by equivalence classes of coordinate triples $[u, v, w]$, where $u, v, w \in GF(4)$. Choose a primitive element t in $GF(4)$ and define three permutations of $PG(2, 4)$ as follows:

$$f_1[u, v, w] = [u^2 + vw, v^2, w^2],$$
$$f_2[u, v, w] = [u^2, v^2, w^2t],$$
$$f_3[u, v, w] = [u^2, v^2, w^2].$$

These are all permutations of the points of PG(2, 4), since $u \mapsto u^2$ is an automorphism of the field GF(4).

3.6.1 **Theorem.** <u>The permutation group</u> PSL(3, 4) <u>acting on</u> PG(2, 4) <u>has a one-point extension.</u>

Proof. We use the notation of Theorem 1.5.2, with $G = $ PSL(3, 4) and $X = $ PG(2, 4), and $* = \infty$. Let h switch ∞ and [1, 0, 0], and act like f_1 on the rest of X. Let g be defined by $g[u, v, w] = [v, u, w]$; the matrix of g has determinant $-1 = 1$ and so g is in G. The points x and y of 1.5.2 are [1, 0, 0] and [0, 1, 0]. We have to check conditions (iii) and (iv) of that theorem.

Clearly, $g^2 = 1$; and $(gh)^3 = 1$ also, as may be proved by the following calculation:

$$(gh)[u, v, w] = [v^2, u^2 + vw, w^2],$$
$$(gh)^2[u, v, w] = [u + v^2w^2, v + (u^2 + vw)w^2, w],$$
$$(gh)^3[u, v, w] = [uw + v^2(w^3 + 1), vw + u^2(w^3 + 1), w^2],$$

provided $[u, v, w] \neq [1, 0, 0]$ or $[0, 1, 0]$. Now if $w \neq 0$, then $w^3 = 1$ and the right-hand side is $[uw, vw, w^2] = [u, v, w]$. If $w = 0$ and $uv \neq 0$, the right-hand side is $[v^2, u^2, 0] = [u, v, 0]$, as required. Finally, the points [1, 0, 0], [0, 1, 0], and ∞ may be dealt with separately. Thus condition (iii) of 1.5.2 is satisfied.

For condition (iv), suppose $p \in G_x$; that is, p is an element of PSL(3, 4) fixing [1, 0, 0], and so it has a representative matrix

$$P = \begin{bmatrix} 1 & a & l \\ 0 & b & m \\ 0 & c & n \end{bmatrix}, \quad \det P = bn - cm = 1.$$

A computation shows that hph may be represented by the matrix

$$\begin{bmatrix} 1 & a^2 + bc & \ell^2 + mn \\ 0 & b^2 & m^2 \\ 0 & c^2 & n^2 \end{bmatrix} \quad ,$$

which fixes $[1, 0, 0]$ and belongs to $PSL(3, 4)$ since $b^2n^2 - c^2m^2 = (bn - cm)^2 = 1$.

Thus the group $\langle G, h \rangle$ is a one-point extension of $G = PSL(3, 4)$. $/\!\!/$

Since (by 2.5.4) $PSL(3, 4)$ is 2-transitive on the 21 points of $PG(2, 4)$, the extension is 3-transitive on 22 points; it is the Mathieu group M_{22}. The order of M_{22} is $22. |PSL(3, 4)| = 443,520$. As we have already indicated, two more extensions are possible.

3.6.2 Theorem. M_{22} has a one-point extension M_{23}, and M_{23} has a one-point extension M_{24}.

Proof. We sketch the proof, which follows the details of 3.6.1 very closely.

To extend M_{22}, introduce the new symbol ∞', define h' to be the composite of f_2 and the transposition $(\infty \ \infty')$, and take $g' = h$. The conditions of 1.5.2 may be verified, so that $M_{23} = \langle M_{22}, h' \rangle$ is a one-point extension of M_{22}.

Similarly, to extend M_{23}, introduce the new symbol ∞'', define h'' to be the composite of f_3 and the transposition $(\infty' \ \infty'')$, and take $g'' = h'$. This yields the one-point extension $M_{24} = \langle M_{23}, h'' \rangle$. $/\!\!/$

The groups M_{23} and M_{24} are also known as Mathieu groups. They are, respectively, 4-transitive and 5-transitive on 23 and 24 points.

3.6.3 Theorem. The Mathieu groups M_{22}, M_{23}, and M_{24} are all simple.

Proof. The stabilizer of a point in M_{22} is the simple group $PSL(3, 4)$, and since M_{22} is 3-transitive on 22 points it can have no regular normal subgroup, by 1.7.6(ii). Thus Theorem 1.6.7 tells us that M_{22} is simple. A parallel argument works for M_{23} (stabilizer

M_{22}) and M_{24} (stabilizer M_{23}). //

We now turn to the construction of designs associated with these groups, following the method explained in the previous section. We begin with the 2 - (21, 5, 1) design PG(2, 4), and the group PSL(3, 4) acting 2-transitively on its points and transitively on its blocks. We have seen that there is a one-point extension M_{22}. Thus, the flow diagram procedure leads to a 3 - (22, 6, λ^+) design, and (by 3.5.5) this is an extension of PG(2, 4) if $\lambda^+ = 1$.

3.6.4 Theorem. PG(2, 4) is extendable, giving a 3 - (22, 6, 1) design, on which M_{22} acts as a group of automorphisms.

Proof. In the notation of Theorem 3.4.3 take β to be the union of ∞ with the line l in PG(2, 4) whose equation is $w = 0$. Let $\mathcal{B}^+$ denote the set of blocks of the design generated by the action of M_{22} on β, and let $b^+ = |\mathcal{B}^+|$.

The setwise stabilizer of β in M_{22} is transitive on the six points of β, since it contains h (which switches ∞ and [1, 0, 0]) and the setwise stabilizer of l in PSL(3, 4) (which is transitive on the five points of l). Thus we may calculate as follows, using the abbreviations $H = M_{22}$, $G = PSL(3, 4)$:

$$
\begin{aligned}
b^+ &= |H : H_{(\beta)}| \\
&= |H : G_{(l)}| / |H_{(\beta)} : G_{(l)}| \\
&= |H : G| \cdot |G : G_{(l)}| / |H_{(\beta)}^{(\infty)}| \\
&= 22.21/6 = 77.
\end{aligned}
$$

So we have $r^+ = 21$, $\lambda_2^+ = 5$, $\lambda^+ = 1$, and by Theorem 3.5.5, the new design is an extension of the original one. //

In the same way, we may use the fact that M_{22} has a one-point extension M_{23} to construct a 4 - (23, 7, 1) design, and the fact that M_{23} has a one-point extension M_{24} to construct a 5 - (24, 8, 1) design. It seems that this family of groups and designs is the result of a fortuitous concatenation of circumstances, numerical and group-theoretical.

The smaller Mathieu groups M_{10}, M_{11}, and M_{12} (Section 1.9) also induce designs. In this case, we begin with the 2 - (9, 3, 1) design constructed at the end of Section 3.4, and it is necessary to choose the group of automorphisms with some care. Not only does the design admit the affine group A of the field GF(9) as a group of automorphisms, it admits the affine group T of the near-field N(2, 3) as well. Both groups act sharply 2-transitively on the same point-set (compare 2.8.5), and both are subgroups of the full group of automorphisms of the design. In fact, both groups have one-point extensions, A^+ and T^+, and these both yield the same 3 - (10, 4, 1) design. Once again, the full group of automorphisms contains A^+ and T^+. However, at the next step there is only one choice. A^+ has no one-point extension, whereas T^+ is just M_{10} (2.8.6) and so it can be extended twice more, giving M_{11} and M_{12}. The standard method yields associated designs, with parameters 4 - (11, 5, 1) and 5 - (12, 6, 1).

We conclude with some remarks on the transitivity problem. The foregoing constructions might tend to give the impression that a t-design (at least for $t > 3$) must admit a group of automorphisms acting t-transitively on the points. This is not so, although it is only recently that counter-examples have been found (see 3.8). However, it is still true that some individual sets of parameters do force the existence of an associated group. For example, we have seen (2.9) that there is a unique 2 - (21, 5, 1) design: it is PG(2, 4) and must therefore admit the 2-transitive group PSL(3, 4).

3.7   Project: Hadamard matrices and designs

An $n \times n$ matrix $H = (h_{ij})$, with $h_{ij} = \pm 1$, is said to be a Hadamard matrix of order n if $HH^t = nI$. It is easy to see that n must be 1, 2, or a multiple of 4. A Hadamard matrix is normalized if the entries in the first row and column are all 1. Clearly every Hadamard matrix can be transformed into a normalized one by multiplying some rows and columns by -1.

3.7.1   Let H be a normalized Hadamard matrix of order $n = 4\lambda + 4$. Delete the first row and column of H, and replace -1 by 0

74

in what remains. Show that the resulting matrix A is the incidence matrix of a symmetric $2 - (4\lambda + 3, 2\lambda + 1, \lambda)$ design. (A design with these parameters is called a <u>Hadamard design.</u>)

3.7.2   Show that the converse of the preceding result is also true.

3.7.3   Construct a Hadamard matrix of order 12. (Use 3.7.2 and a suitable design.)

3.7.4   Suppose that $(X, \mathcal{B})$ is a Hadamard design, and ∞ is a new point not in X. Show that the sets of the form $\beta \cup \{\infty\}$ and $X - \beta$ $(\beta \in \mathcal{B})$ are the blocks of a $3 - (4\lambda + 4, 2\lambda + 2, \lambda)$ design, and that this is an extension of $(X, \mathcal{B})$.

3.7.5   Prove that any extension of a Hadamard design is isomorphic to the extension constructed in 3.7.4. (Show that two intersecting blocks in an extension must meet in exactly $\lambda + 1$ points, and conclude that the complement of each block in the extension is also a block.)

3.7.6   Let Δ_{11} denote the $2 - (11, 5, 2)$ design given at the beginning of Section 3.2, and let Δ_{12} denote its extension constructed by the method of 3.7.4. (We aim to show that the group of automorphisms of Δ_{12} is the simple group M_{11}.) Let G be the automorphism group of Δ_{11}: show that G acts 2-transitively on the points and $|G_{01}| = 6$.

3.7.7   Show that G_0 is primitive in its action on the ten points $1, 2, \ldots, T$, and that G_0 is simple. Deduce that G is simple. (In fact, $G_0 \approx A_5$ and $G \approx PSL(2, 11)$, but we do not need this.)

3.7.8   Let H be the automorphism group of Δ_{12}. Show that $H_\infty = G$, and that H acts 3-transitively on the points. Deduce that H is simple.

3.7.9*   Show that H acts 4-transitively on the 11 pairs of complementary blocks in Δ_{12}. Identify H with the group M_{11} constructed in 1.9.8. (We have now proved that M_{11} is simple, a fact assumed in 1.9.11.)

3.8 Project: A 5 - (28, 7, 1) design

In recent years several new 5-designs have been constructed, and some of them have no associated 5-transitive group. In these examples the design property is a combinatorial accident, not forced by transitivity. In this project we introduce a 5 - (28, 7, 1) design due to R. H. F. Denniston.

3.8.1 Let $G = \text{PSL}(2, 27)$ act on the projective line $L = \text{GF}(27) \cup \{\infty\}$, as in Section 2.5. Show that G acts 2-transitively but not 3-transitively on L. Check that G is transitive on the <u>unordered</u> 3-subsets of L.

3.8.2 Let t be a primitive element of $\text{GF}(27)$ satisfying $t^3 = t+2$, and let β_1, β_2 be the following 7-subsets of L:

$$\beta_1 = \{\infty,\ 0,\ 1,\ 2,\ t,\ t+1,\ t+2\},$$
$$\beta_2 = \{\infty,\ 0,\ 1,\ 2t^2 + 2t,\ 2t^2,\ t^2 + 2t + 2,\ t^2 + 2t\}.$$

Show that the setwise stabilizers in G of β_1 and β_2 have orders 3 and 7 respectively. (For β_2, consider $u \mapsto -(t^2 + 2t)/(u - (t^2 + 2t))$.)

3.8.3 Show that the construction of Theorem 3.4.3, applied to β_1 and β_2 respectively, leads to a 3 - (28, 7, 35) design and a 3 - (28, 7, 15) design.

3.8.4* Show that the union of the two designs has the correct number of blocks for a 5 - (28, 7, 1) design. Verify that the union <u>is</u> a 5-design by considering the 50 blocks containing ∞, 0, and 1. (Do not attempt the last part if time is short!)

3.8.5* Show that this 5 - (28, 7, 1) design does not admit a group of automorphisms acting 5-transitively on the points.

3.9 Project: Uniqueness of the 3 - (22, 6, 1) design

The contraction of a 3 - (22, 6, 1) design is a 2 - (21, 5, 1) design, and so (by 2.9) it must be the unique projective plane $PG(2, 4)$

of order 4. We shall use this fact to show that the 3 - (22, 6, 1)
design itself is unique (up to isomorphism). Remarkably, the existence
of the 5 - (24, 8, 1) design acted on by M_{24} also plays an important
part in the proof.

3.9.1 Let (X, $\mathcal{B}$) be a 3 - (22, 6, 1) design. Choose x in X,
and label the points of X - {x}, and the blocks of the contraction with
respect to x, so as to correspond with PG(2, 4). (This can be done,
by 2.9.) Show that a block of $\mathcal{B}$ not containing x corresponds to a
hyperoval in PG(2, 4).

3.9.2 Let us use the term 'extension class' for a family $\mathcal{E}$ of
6-subsets of the points of PG(2, 4), such that the lines of PG(2, 4) (each
augmented by a new point x) and the members of $\mathcal{E}$ are the blocks of a
3 - (22, 6, 1) design. Show that $\mathcal{E}$ consists of 56 hyperovals in
PG(2, 4), with the property that any two distinct members of $\mathcal{E}$ intersect
in 0 or 2 points.

3.9.3 Show that PG(2, 4) contains exactly 168 hyperovals.

3.9.4 Let (Z, $\mathcal{L}$) be the 5 - (24, 8, 1) design constructed in
3.6, and let a, b, c $\in$ Z. Identify PG(2, 4) with the triple contraction
of (Z, $\mathcal{L}$) with respect to a, b, and c. Show that there are 56 members
of $\mathcal{L}$ containing a and b, but not c, and that these correspond to an
'extension class'.

3.9.5 Deduce that the relation of 'intersecting in 0, 2 or 6 points'
is an equivalence relation on the set of 168 hyperovals, and that the
equivalence classes are just the 'extension classes'.

3.9.6 Show that the setwise stabilizer of {a, b, c} in the action
of M_{24} on the points of (Z, $\mathcal{L}$) is transitive on {a, b, c}. Deduce
that the extensions of PG(2, 4) constructed using the three possible
'extension classes' are all isomorphic.

Questions related to designs were studied by Steiner, Kirkman, E. H. Moore, and others in the nineteenth century. In the 1930s, Fisher and Yates introduced the crucial notion of a 'balanced' design, corresponding to what is now called a 2-design. Fisher and Yates were motivated by practical considerations connected with the design of experiments. The mathematics of designs was developed in papers by Bose (1939) and Bruck and Ryser (1949).

The standard works of reference are by Dembowski [2] and Hall [4]. (The useful table of 2-designs in [4] is updated in [5].) Interest in the connections between groups and t-designs was stimulated by the paper of Hughes [7]. The paper of Denniston [3] contains details of the construction of many 5-designs, including the one discussed in Section 3. 8.

The Mathieu groups M_{22}, M_{23}, M_{24} and the associated designs were investigated by Witt in two classic papers [9], [10]. Further developments, including uniqueness proofs, may be found in the books of Lüneburg [8] and Cameron [1].

(Readers unfamiliar with the number-theoretic results quoted in Section 3. 3 should consult Hardy and Wright [6], or a similar text.)

1.   Cameron, P. J. Parallelisms of Complete Designs. (Cambridge University Press, 1976.)

2.   Dembowski, P. Finite Geometries. (Springer-Verlag, 1968.)

3.   Denniston, R. H. F. Some new 5-designs. Bull. London Math. Soc. , 8 (1976), 263-7.

4.   Hall, M. Combinatorial Theory. (Blaisdell, 1967.)

5.   Hall, M. Construction of block designs. A Survey of Combinatorial Theory (J. N. Srivastava, ed.). (North-Holland, 1973.)

6.   Hardy, G. H. and Wright, E. M. The Theory of Numbers. (Oxford University Press, 1960.)

7.   Hughes, D. R. On t-designs and groups. Amer. J. Math. , 87 (1965), 761-8.

8.   Lüneburg, H. Transitive Erweiterungen endlicher Permutationsgruppen. (Springer-Verlag, 1969.)

9. Witt, E. Die 5-fach transitiven Gruppen von Mathieu. _Abh._
    Math. Sem. Univ. Hamburg, 12 (1938), 256-64.

10. Witt, E. Uber Steinersche Systeme. _Abh. Math. Sem. Univ._
    Hamburg, 12 (1938), 265-75.

4 · Groups and Graphs

'The method of graphs displays, by virtue of its highly geometrical
interpretation (which leads to the solution of many of the following
questions), the equivalence of apparently remote researches. '
          D. König, in the introduction to his paper <u>Über Graphen</u>
          <u>und ihre Anwendung auf Determinantentheorie und</u>
          <u>Mengenlehre</u>, 1916.

4.1 Permutation groups and graphs

In this chapter we shall be concerned with the relationship between
permutation groups and graphs. We begin by explaining how a transitive
permutation group may be represented graphically, and then we reverse
the process, showing that a graph gives rise to a permutation group.
Some very interesting groups may be constructed in this way - in
particular, a sporadic simple group discovered by Higman and Sims in
1967.

Let us suppose that a transitive permutation group (G, X) is
given; then there is an induced action of G on $X \times X$, defined by

$$g(x, y) = (gx, gy).$$

Since G is transitive, the diagonal $\Delta = \{(x, x) | x \in X\}$ is an orbit. In
order to standardize the notation, we shall write D_0 instead of Δ for
the rest of this chapter.

    4.1.1 Lemma. <u>Let (G, X) be a transitive permutation group,
and suppose that G has r orbits $D_0, D_1, \ldots, D_{r-1}$ on $X \times X$. Then
r is the rank of</u> (G, X).

    Proof. Fix $x \in X$, and put

$$D_i(x) = \{y \in X | (x, y) \in D_i\} \quad (i = 0, 1, \ldots, r-1).$$

80

It is easy to verify that $\{x\} = D_0(x)$, $D_1(x)$, $\ldots$, $D_{r-1}(x)$ are precisely the orbits of G_x on X. Hence the result follows. //

Each orbit D_i $(0 \le i \le r-1)$, as a subset of $X \times X$, defines a relation on X. This relation may be represented graphically in the following way. The elements of X are represented by points (called 'vertices'), and two points x and y are joined by a directed line segment ('arc') from x to y whenever $(x, y) \in D_i$. This gives a <u>directed graph</u>, or <u>digraph</u>, $\Gamma(D_i)$ associated with each orbit $D_i \neq D_0$. For example, if G is the group generated by (12345), acting regularly on the set $X = \{1, 2, 3, 4, 5\}$, and if D_1 is the orbit containing (1, 2), then the digraph $\Gamma(D_1)$ is as follows (Fig. 3):

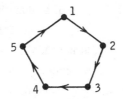

Fig. 3

In abstract terms, the digraph $\Gamma(D_i)$ is logically identical with the relation D_i on X, but the diagrammatic representation can be a useful aid to intuition. In this spirit, we shall obtain a characterization of primitivity in terms of the associated digraphs.

4.1.2 Definition. A <u>walk</u> of <u>length</u> r in a digraph Γ is a sequence $(v_0, v_1, \ldots, v_r)$ of vertices of Γ such that, for $i = 1, 2, \ldots, r$, either (v_{i-1}, v_i) or (v_i, v_{i-1}) is an arc of Γ. The digraph Γ is . <u>connected</u> if, given any two vertices u and w, there is a walk beginning at u and ending at w.

4.1.3 Theorem. <u>Let</u> (G, X) <u>be a transitive permutation group.</u> G <u>acts primitively on</u> X <u>if and only if the digraph</u> $\Gamma(D)$ <u>is connected,</u> <u>for each orbit</u> D $(\neq D_0)$ <u>of</u> G <u>on</u> $X \times X$.

Proof. Suppose that (G, X) is primitive, and let D $(\neq D_0)$ be an orbit of G on $X \times X$. Let R denote the relation on X defined by

81

(x, y) $\in$ R if and only if x and y may be joined by a walk in $\Gamma(D)$. R is G-admissible, since if $x = v_0, v_1, \ldots, v_r = y$ are the vertices of a walk joining x and y, then $gx = gv_0, gv_1, \ldots, gv_r = gy$ are the vertices of a walk joining gx and gy. Now, since G acts primitively, R must be trivial. But $D \neq D_0$ implies that $R \neq D_0$; hence we must have $R = X \times X$. This means that any two vertices of $\Gamma(D)$ may be joined by a walk in $\Gamma(D)$, so that $\Gamma(D)$ is connected.

Conversely, suppose that G acts imprimitively and R is a non-trivial G-admissible equivalence relation on X. Since $R \neq D_0$ we may choose distinct points s and t such that $(s, t) \in R$. Let D $(\neq D_0)$ be the orbit of G on $X \times X$ containing (s, t); since D is an orbit and R is G-admissible, it follows that R contains D. Now, since $R \neq X \times X$ we may choose points w and z such that $(w, z) \notin R$. If there were a walk $w = v_0, v_1, \ldots, v_r = z$ in $\Gamma(D)$, then we should have $(v_{i-1}, v_i) \in R$ for $1 \leq i \leq r$, since $R \supseteq D$ and R is symmetric. But then, by the transitivity of R, we should have $(w, z) \in R$. This contradiction shows that $\Gamma(D)$ is disconnected. $/\!/$

Each orbit D $(\neq D_0)$ is 'paired' with its transpose:

$$D^t = \{(x, y) \in X \times X \,|\, (y, x) \in D\}.$$

In general, D and D^t will be different, as (for example) with the orbit whose digraph is illustrated in Fig. 3. When $D = D^t$, we say that D is 'self-paired', or symmetric. This case often occurs in practice, as indicated by the following result.

4.1.4 Theorem. Let (G, X) be a transitive permutation group and suppose $|G|$ is even. Then at least one orbit D $(\neq D_0)$ of G on $X \times X$ is symmetric.

Proof. Since $|G|$ is even, G contains an element g of order 2, which must switch some pair x, y $(x \neq y)$. Let D be the orbit containing (x, y). D contains (y, x) also, since $g(x, y) = (y, x)$, and so $D = D^t$. $/\!/$

When D is symmetric, the graphical representation may be simplified. We may replace the arc from x to y and the arc from y to x by a single, undirected line. We now have a 'graph', whose 'vertices' are the members of X, and whose 'edges' are pairs $\{x, y\}$ with $(x, y) \in D$. At this point, a formal definition seems desirable.

4.1.5 Definition: A graph Γ is a pair (V, E), where V is a finite set whose members are called vertices, and E is a subset of the set $V^{(2)}$ of unordered pairs of vertices. The members of E are called edges. (Occasionally we use the notation $V\Gamma$ and $E\Gamma$ for V and E.) If $\{v, w\}$ is an edge of Γ, we say that the vertices v and w are adjacent.

We emphasize that a graph is an abstract concept. We frequently use plane diagrams of points (representing vertices) and lines (representing edges) to illuminate the structure of a graph, but the definition is completely independent of such notions. When we use the plane diagrams to represent graphs, we must be careful to indicate that some 'crossings' of the lines representing edges are merely due to deficiencies of the representation, and do not represent vertices. For example, the complete graph K_n has $V = \{1, 2, \ldots, n\}$ and $E = V^{(2)}$; each pair of distinct vertices is joined by an edge. K_4 may be represented without crossings, but for K_5 they are essential (Fig. 4). (In Chapter 5 we shall allow representations of graphs on surfaces other than the plane, thereby avoiding the need for crossings altogether.)

Fig. 4

The concepts introduced in Definition 4.1.2 may be carried over to graphs in the obvious way. A walk of length r is a sequence $(v_0, v_1, \ldots, v_r)$ such that $\{v_{i-1}, v_i\}$ is an edge for $1 \le i \le r-1$. Γ is connected if any two vertices can be joined by a walk.

Let us now return to the graphical representation of permutation groups. A simple example is provided by the action of the dihedral group $G = \langle (1234),\ (13) \rangle$ on the set $X = \{1,\ 2,\ 3,\ 4\}$. As we noted in Section 1.3, the rank is 3 and the orbits on $X \times X$ are:

$$D_0 = \{(1,\ 1),\ (2,\ 2),\ (3,\ 3),\ (4,\ 4)\},$$
$$D_1 = \{(1,\ 3),\ (2,\ 4),\ (3,\ 1),\ (4,\ 2)\},$$
$$D_2 = \{(1,\ 2),\ (1,\ 4),\ (2,\ 3),\ (3,\ 4),\ (2,\ 1),\ (4,\ 1),\ (3,\ 2),\ (4,\ 3)\}.$$

Both D_1 and D_2 are symmetric and so we obtain two graphs, as in Fig. 5. We remark that $\Gamma(D_1)$ is disconnected, as we should expect,

$$\Gamma(D_1) \qquad\qquad\qquad \Gamma(D_2)$$

Fig. 5

since G acts imprimitively. A more substantial example is provided by the action of S_n on the set $X = N^{(2)}$, where $N = \{1,\ 2,\ \ldots,\ n\}$. (Note that the vertices of the graph are themselves pairs, so the edges will represent pairs of pairs.) We use the abbreviation ij for $\{i,\ j\} \in X$, so that $ij = ji$. S_n is transitive on X, and it has rank 3; the orbits are

D_0, comprising pairs (ij, ij);

D_1, comprising pairs (ij, kl), with i, j, k, l all different;

D_2, comprising pairs (ij, il), with i, j, l all different.

Again, $D_1^t = D_1$ and $D_2^t = D_2$, so that we have two 'complementary' graphs. In the case $n = 5$, and using the orbit D_1, we obtain the famous Petersen graph (Fig. 6).

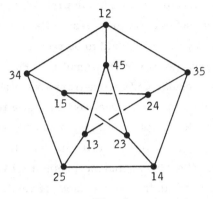

12

45

34

35

15   24

13   23

25   14

Fig. 6

4.2   Automorphisms of graphs

4.2.1 **Definition.**   An <u>automorphism</u> of a graph Γ with vertex-set V and edge-set E is a permutation α of V such that $\{v, w\} \in E$ implies that $\{\alpha v, \alpha w\} \in E$.

The set of all automorphisms forms a permutation group G acting on V; we shall refer to this as the (full) <u>automorphism group</u> of Γ, and denote it by $\text{Aut } \Gamma$. Clearly, similar concepts apply to digraphs, but we shall not need that generalization.

The full group of automorphisms of the complete graph K_n is the symmetric group S_n, since in this case any permutation of the vertices preserves adjacency. Any subgroup of S_n (that is, any permutation group of degree n) is also a group of automorphisms of K_n. The only other graph on n vertices which admits S_n as a group of automorphisms is the 'complement' of K_n - the graph with n vertices and no edges. (In general, the <u>complement</u> of a graph Γ is the graph Γ^c with the same set of vertices and edge-set $E^c = V^{(2)}$ - E. Clearly, the automorphism groups of Γ and Γ^c are the same.)

4.2.2 **Definition.**   A graph is <u>vertex-transitive</u> if it admits a group of automorphisms acting transitively on the vertices.

In a vertex-transitive graph all vertices have the same properties
with respect to the structure of the graph. In particular, each vertex
is contained in the same number (κ) of edges: κ is called the <u>valency</u>
of the vertex, and the graph itself is said to be <u>regular</u>, or κ-<u>valent.</u> A
regular graph may or may not be vertex-transitive. For example, the
first graph in Fig. 7 is regular and vertex-transitive, whereas the second
is regular but not vertex-transitive. One way of proving this is by
examining the circuits in the graph: a <u>circuit</u> is an ordered set of
distinct vertices (a, b, c, ..., f) such that {a, b}, {b, c}, ..., {f, a}
are edges of the graph. Clearly, an automorphism must transform a
circuit into another circuit with the same number of vertices. In Fig. 7(b)
the vertex v lies in a 3-circuit (or triangle) but w does not, and so no
automorphism can take v to w.

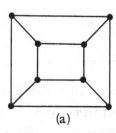
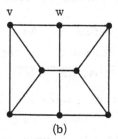

(a)   (b)

Fig. 7

The relationship between the automorphism group of a vertex-
transitive graph and the graph associated with a transitive group is
elucidated in the following theorem.

4.2.3 Theorem. (i) <u>Let</u> (G, X) <u>be a transitive permutation</u>
<u>group and</u> D <u>a symmetric orbit of</u> G <u>on</u> X × X. <u>Then the associated</u>
<u>graph</u> $\Gamma(D)$ <u>admits</u> G <u>as a group of automorphisms.</u>

(ii) <u>Let</u> $\Gamma = (V, E)$ <u>be a graph and suppose that</u> G <u>is a group</u>
<u>of automorphisms of</u> Γ <u>acting transitively on</u> V; <u>then</u> E <u>is a union of</u>
<u>orbits of</u> G <u>on</u> V × V. <u>Precisely, there is a family</u> $\mathfrak{D}$ <u>of orbits such</u>
<u>that</u>

$$\{u, v\} \in E \Longleftrightarrow (u, v) \in D \quad \text{for some } D \in \mathfrak{D}.$$

Proof. (i) If $\{u, v\}$ is an edge of $\Gamma(D)$, then, by definition, $(u, v) \in D$. Since D is an orbit, $(gu, gv) \in D$ for any g in G, and so $\{gu, gv\}$ is an edge. Thus G acts as a group of automorphisms.

(ii) Suppose Γ and G are as stated. Define the family $\mathfrak{D}$ of orbits by the rule that D belongs to $\mathfrak{D}$ if and only if there is some edge $\{x, y\}$ with $(x, y) \in D$. The definition ensures that the left-to-right implication above holds. Conversely, if $(u, v) \in D \in \mathfrak{D}$, then there is some edge $\{x, y\}$ with $(x, y) \in D$. Since D is an orbit, $u = gx$ and $v = gy$ for some g in G; and since g is an automorphism, $\{u, v\}$ is an edge. $/\!/$

4.3 Rank 3 groups and the associated graphs

Suppose that (G, X) is transitive, of rank three, with orbits D_0, D_1, D_2 of G on $X \times X$. (As usual, D_0 denotes the diagonal.) If D_1 is symmetric, then D_2 must be symmetric as well, and so the associated objects $\Gamma(D_1)$ and $\Gamma(D_2)$ are both graphs, rather than digraphs. Since D_1 and D_2 are the only non-trivial orbits, the two graphs must be complementary; further, if G acts primitively on X, both graphs are connected (4.1.3).

We shall show that the two complementary graphs have some special combinatorial properties. Clearly, both graphs are regular, since (4.2.3) G acts as vertex-transitive group of automorphisms.

4.3.1 Theorem. Let (G, X) be a transitive group of rank 3. Suppose that $D \ (\neq D_0)$ is a symmetric orbit of G on $X \times X$ and $\Gamma = \Gamma(D)$ is the associated graph. Then

(i) Given any two adjacent vertices of Γ, there is a constant number (say α) of vertices adjacent to both of them.

(ii) Given any two distinct non-adjacent vertices of Γ, there is a constant number (say γ) of vertices adjacent to both of them.

Proof. (i) For any two elements x, y of X define

$$N(x, y) = \{z \in X \,|\, (x, z) \in D \text{ and } (y, z) \in D\}.$$

In other words, $N(x, y)$ is the set of vertices of Γ which are adjacent to both x and y. If x and y are adjacent, then $(x, y) \in D$. Similarly, if x' and y' are adjacent we have $(x', y') \in D$ and, since D is an orbit, there is some g in G such that $gx = x'$ and $gy = y'$. The mapping $z \mapsto gz$ is a one-to-one correspondence between $N(x, y)$ and $N(x', y')$, so that $|N(x, y)| = |N(x', y')| = \alpha$, as required.

(ii)   If x and y are two distinct non-adjacent vertices, then (since there are only two non-trivial orbits) the pair (x, y) belongs to the other non-trivial orbit, $(x, y) \in \overline{D} \neq D$. Any other distinct non-adjacent pair (x', y') also belongs to $\overline{D}$, and so there is some $\overline{g}$ in G such that $\overline{g}x = x'$, $\overline{g}y = y'$. Repeating the argument as in (i), with $\overline{g}$ instead of g, gives $|N(x, y)| = |N(x', y')| = \gamma$, as required. //

4.3.2 **Definition.**   A regular graph of valency κ (not a complete graph or its complement), in which the conditions (i) and (ii) of Theorem 4.3.1 hold, is said to be <u>strongly regular,</u> with parameters (κ, α, γ).

The combinatorial structure of a strongly regular graph may be indicated diagrammatically as in Fig. 8.

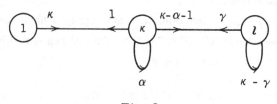

Fig. 8

The diagram indicates that if we select one vertex (the left-hand circle), it is joined to κ others (the central circle). Each of these κ vertices is joined to the initial vertex, to α vertices adjacent to the initial one, and to $\kappa - \alpha - 1$ others (since every vertex must have valency κ). As for the remaining l vertices (the right-hand circle), each is adjacent to γ vertices in the central circle and to $\kappa - \gamma$ vertices in its own circle. Applying the basic counting principle (1.2.2) to the set of edges joining the central circle to the right-hand one, we find

4.3.3 $\gamma l = \kappa(\kappa - \alpha - 1)$.

Theorem 4.3.1 tells us that the graph of a rank 3 group, with respect to a symmetric orbit, is strongly regular. For example, Petersen's graph (Fig. 6) is strongly regular, with parameters $(3, 0, 1)$. In general, the action of S_n on $N^{(2)}$, with respect to the orbit D_1 as defined in Section 4.1, yields a strongly regular graph with $\kappa = \frac{1}{2}n(n - 1)$, $\alpha = \frac{1}{2}(n - 4)(n - 5)$, $\gamma = \frac{1}{2}(n - 3)(n - 4)$. The other orbit D_2 gives rise to a complementary graph.

If we are given a strongly regular graph, then it may or may not admit a group of automorphisms acting transitively and with rank 3 on the vertices. Let us suppose that we are given a graph Γ and we know that it does admit such a group of automorphisms G. Then the two non-trivial orbits D_1 and D_2 must consist precisely of the pairs of adjacent vertices, and the pairs of distinct non-adjacent vertices; the associated graph $\Gamma(D_1)$ is just Γ, and $\Gamma(D_2)$ is the complementary graph Γ^c. Thus, by 4.3.1, both graphs are strongly regular. If Γ has parameters (κ, α, γ) and l is defined by 4.3.3, then the parameters of Γ^c are

$$\bar{\kappa} = l, \quad \bar{\alpha} = l - \kappa + \gamma - 1, \quad \bar{\gamma} = l - \kappa + \alpha + 1.$$

We shall need a criterion for G to act primitively on the vertices of Γ, and Theorem 4.1.3 tells us that it is sufficient that both Γ and Γ^c should be connected. This means that the parameters γ and $\bar{\gamma}$ must both be non-zero. If we are given that Γ is connected, then the condition $\bar{\gamma} = 0$ is equivalent to $\gamma = \kappa$. Thus we may summarise these remarks in the following useful form.

4.3.4 Theorem. <u>Let</u> Γ <u>be a connected graph which admits a rank</u> 3 <u>group of automorphisms</u> G. <u>Then</u> G <u>acts primitively on the vertices if and only if the parameters</u> γ <u>and</u> κ <u>of</u> Γ <u>(considered as a strongly regular graph) are not equal.</u> //

4.4 Feasibility conditions for strongly regular graphs

It is clear that not every possible set of parameters (κ, α, γ) can be associated with a strongly regular graph; for example, if the number l

defined by 4.3.3 is not an integer then there cannot be a graph. This simple condition shows that parameters like (12, 0, 5) and (11, 3, 6) are 'not feasible'. In this section we shall obtain much more powerful feasibility conditions, based on simple techniques from linear algebra.

4.4.1 **Definition.** Let Γ be a graph with n vertices. The adjacency matrix of Γ is the $n \times n$ matrix A whose rows and columns are labelled by the vertices of Γ, and whose entries are given by

$$(A)_{uv} = \begin{cases} 1 & \text{if } u \text{ and } v \text{ are adjacent in } \Gamma; \\ 0 & \text{otherwise.} \end{cases}$$

4.4.2 **Lemma.** Suppose that A is the adjacency matrix of a graph Γ, and A^l denotes the lth power of A $(l \geq 0)$. Then $(A^l)_{uv}$ is equal to the number of walks of length l in Γ, beginning at u and ending at v.

Proof. We proceed by induction on l. The statement is clearly true when $l = 0$ $(A^0 = I)$, and when $l = 1$ $(A^1 = A)$.

Suppose that the result is true for $l - 1$. A walk of length l from the vertex u to the vertex v may be uniquely decomposed into a walk of length $l - 1$ from u to some vertex w adjacent to v, together with the edge $\{w, v\}$. By the induction hypothesis, the number of walks of length $l - 1$ from u to w is $(A^{l-1})_{uw}$. Thus the number of walks of length l from u to v is

$$\sum_{\{w, v\} \in E} (A^{l-1})_{uw} = \sum_{w \in V} (A^{l-1})_{uw}(A)_{wv} = (A^l)_{uv}.$$

Hence the result is true for all $l \geq 0$. //

4.4.3 **Theorem.** Let Γ be a strongly regular graph with parameters (κ, α, γ); then the adjacency matrix A of Γ satisfies the equation

$$A^2 + (\gamma + \alpha)A + (\gamma - \kappa)I = \gamma J.$$

Proof. We consider the terms $(A^i)_{uv}$ for i = 0, 1, 2. There are three cases for each i, according as u = v, u and v are adjacent, or u and v are distinct and not adjacent.

Suppose $i = 2$. The number $(A^2)_{uv}$ is just the number of walks of length 2 from u to v. If $u = v$, there are κ such walks (u, x, u), where x is any one of the κ vertices adjacent to u. If u and v are adjacent, there are α such walks (u, y, v), where y is any one of the α vertices adjacent to both u and v. If u and v are distinct and not adjacent, there are γ such walks (u, z, v), where z is any one of the γ vertices adjacent to both u and v.

Similar arguments when $i = 1$ and $i = 0$ lead to the following table:

	$(A^2)_{uv}$	$(A)_{uv}$	$(I)_{uv}$
$u = v$	κ	0	1
$\{u, v\} \in E$	α	1	0
$\{u, v\} \notin E$	γ	0	0 .

It follows that in all cases $[A^2 + (\gamma - \alpha)A + (\gamma - \kappa)I]_{uv} = \gamma$, as claimed. //

We now transform this algebraic condition into a numerical one, involving only the parameters (κ, α, γ).

4.4.4 Theorem. If a strongly regular graph Γ with parameters (κ, α, γ) exists, then either

(i) $\kappa = 2\gamma$ and $\alpha = \gamma - 1$, or

(ii) $(\alpha - \gamma)^2 + 4(\kappa - \gamma)$ is a perfect square, say s^2, and the expression

$$m = (\kappa/2\gamma s)[(\kappa - 1 + \gamma - \alpha)(s + \gamma - \alpha) - 2\gamma]$$

is a positive integer.

Proof. Let n be the number of vertices of Γ, so that A is an $n \times n$ matrix, and let j denote the column vector whose n entries are all equal to 1. Since A has κ ones in each row, $Aj = \kappa j$; that is, j is an eigenvector of A associated with the eigenvalue κ.

Suppose that e is any eigenvector of A, with associated eigenvalue λ. The equation $A^2 + (\gamma - \alpha)A + (\gamma - \kappa)I = \gamma J$ shows that e is also an eigenvector of J, and the associated eigenvalue is $\lambda^2 + (\gamma - \alpha)\lambda + (\gamma - \kappa)$. Now the eigenvectors and eigenvalues of γJ are easily found. j is an eigenvector associated with the eigenvalue γn, which checks with the remarks in the previous paragraph, since $\kappa^2 + (\gamma - \alpha)\kappa + (\gamma - \kappa) = \gamma n$. The rank of γJ is 1, so that its kernel has dimension $n - 1$ and there is an $(n-1)$-dimensional space of eigenvectors associated with the eigenvalue 0. Thus every eigenvalue $\lambda \neq \kappa$ of A corresponds to the zero eigenvalue of γJ, and satisfies

$$\lambda^2 + (\gamma - \alpha)\lambda + (\gamma - \kappa) = 0.$$

We have shown that the eigenvalues of A are κ, λ_1, λ_2, where λ_1, λ_2 are the roots $\frac{1}{2}(\alpha - \gamma \pm \sqrt{d})$ of the quadratic equation, and $d = (\alpha - \gamma)^2 + 4(\kappa - \gamma)$. The multiplicity of κ is unity, and if m_i ($i = 1, 2$) denotes the multiplicity of λ_i we have the equations

$$m_1 + m_2 = n - 1,$$
$$(m_1 - m_2)\sqrt{d} = (m_1 + m_2)(\gamma - \alpha) - 2\kappa.$$

The second equation follows from the fact that the trace of A is zero, so that $\kappa + m_1\lambda_1 + m_2\lambda_2 = 0$.

If $\sqrt{d}$ is irrational, then we deduce that $m_1 = m_2 = \kappa/(\gamma - \alpha)$. Thus we must have $\gamma - \alpha \geq 1$; and if $\gamma - \alpha \geq 2$ then $m_1 + m_2 \leq \kappa$, so that $n - 1 \leq \kappa$ and we have the absurdity $l \leq 0$. Hence, in this case $\gamma - \alpha = 1$, which leads to $\kappa = 2\gamma$ and case (i) of the theorem.

If $\sqrt{d}$ is an integer s, then eliminating m_2 from the two equations leads to the stated expression for $m = m_1$, and this must be an integer since it represents a multiplicity. Thus we have part (ii) of the theorem. //

As an example of the theorem, let us consider the feasibility of the parameters $(\kappa, 0, 1)$. These parameters correspond to a regular graph of valency κ in which there are no 3-circuits (since $\alpha = 0$) and no 4-circuits (since $\gamma = 1$); furthermore, the number of vertices n is $1 + \kappa^2$, and this is the smallest number possible for a κ-valent graph whose shortest circuit has length 5. For this reason, strongly regular graphs with parameters $(\kappa, 0, 1)$ are of interest in themselves. Theorem 4.4.4 tells us that either condition (i) holds and $\kappa = 2$, or (ii) holds, in which case $4\kappa - 3 = s^2$ and

$$m = (\kappa/2s)(\kappa s + \kappa - 2)$$

is an integer. Eliminating κ and writing the result as a polynomial equation in s, we obtain

$$s^5 + s^4 + 6s^3 - 2s^2 + (9 - 32m)s - 15 = 0.$$

Thus s must be a divisor of 15, and the corresponding non-trivial values of $\kappa = (s^2 + 3)/4$ are 3, 7, or 57. In other words, a strongly regular graph with parameters $(\kappa, 0, 1)$ can exist only if $\kappa = 2, 3, 7,$ or 57. It is known that graphs do exist when $\kappa = 2$ (pentagon), $\kappa = 3$ (Petersen's graph), and $\kappa = 7$; in these cases the graphs are unique and each of them admits a rank 3 group of automorphisms. In the case $\kappa = 57$ it is not known whether or not a graph exists, although it has been shown that there cannot be a corresponding rank 3 group.

Another application of Theorem 4. 4. 4 may be found in Section 4. 7.

4. 5   The Higman-Sims group

In this section we shall construct a 'sporadic' simple group, as a group of automorphisms of a strongly regular graph with parameters (22, 0, 6). (It is easy to check that these parameters are feasible.) We shall use the remarkable properties of the unique 2 - (21, 5, 1) design PG(2, 4), and its extension, the unique 3 - (22, 6, 1) design. The relevant properties were established in Sections 2. 9 and 3. 9.

It is convenient to begin by making a slight generalization of the notion of a design. Instead of requiring that a block $\beta \in \mathcal{B}$ should be a set of points $x \in X$, so that the 'incidence relation' is simply the membership relation $x \in \beta$, we shall postulate an arbitrary incidence relation I between arbitrary sets X and $\mathcal{B}$. Thus a $t - (v, k, \lambda)$ design may now be defined as a triple $(X, \mathcal{B}, I)$, with the properties:

(i) X is a v-set;

(ii)   given any 'block' $\beta \in \mathcal{B}$, there are exactly k 'points' $x \in X$ such that $xI\beta$;

(iii) given any t points $x_1, \ldots, x_t$, there are exactly λ
blocks β such that $x_1 I \beta, \ldots, x_t I \beta$.

An automorphism of $(X, \mathcal{B}, I)$ is a permutation h, acting on X and $\mathcal{B}$, such that $xI\beta$ if and only if $hxIh\beta$. It should be clear that this generalization is obtained by merely replacing the ϵ-membership relation by an arbitrary I-incidence relation, and that the basic results in Chapter 3 carry over.

We shall construct the <u>Higman-Sims graph</u> Σ as follows. Let $\mathcal{P}$ be the set of points and $\mathcal{L}$ the set of lines of a projective plane of order 4. As in 3.9.2, let $\mathcal{K}$ be an 'extension class' of 56 hyperovals in $(\mathcal{P}, \mathcal{L})$; each member of $\mathcal{K}$ is a set of 6 points with the property that no three are collinear, and any two members of $\mathcal{K}$ meet in 0 or 2 points. Finally, let 0 and ∞ be two symbols not in $\mathcal{P}$, $\mathcal{L}$, or $\mathcal{K}$. The vertex-set of Σ is

$$V = \{0, \infty\} \cup \mathcal{P} \cup \mathcal{L} \cup \mathcal{K},$$

and its edge-set E is the union of the following seven subsets:

$$E_1 = \{0, \infty\}$$
$$E_2 = \{\{0, p\} \mid p \in \mathcal{P}\}$$
$$E_3 = \{\{\infty, l\} \mid l \in \mathcal{L}\}$$
$$E_4 = \{\{p, l\} \mid p \in \mathcal{P}, l \in \mathcal{L} \text{ and } p \in l\}$$
$$E_5 = \{\{p, H\} \mid p \in \mathcal{P}, H \in \mathcal{K} \text{ and } p \in H\}$$
$$E_6 = \{\{l, H\} \mid l \in \mathcal{L}, H \in \mathcal{K} \text{ and } l \cap H = \phi\}$$
$$E_7 = \{\{H, K\} \mid H, K \in \mathcal{K} \text{ and } H \cap K = \phi\}.$$

The diagram (Fig. 9) is a useful aid to memory. In order to exhibit the symmetry properties of Σ, we shall show that it contains, in a natural way, two 3 - (22, 6, 1) designs.

4.5.1 **Lemma.** <u>Let</u> $X_1 = \mathcal{P} \cup \{\infty\}$, $\mathcal{B}_1 = \mathcal{K} \cup \mathcal{L}$, <u>and define an</u> <u>incidence relation</u> I_1 <u>by the rule that</u>

$$xI_1\beta \Longleftrightarrow \{x, \beta\} \text{ is an edge of } \Sigma .$$

<u>Then</u> $(X_1, \mathcal{B}_1, I_1)$ <u>is a</u> 3 - (22, 6, 1) <u>design.</u>

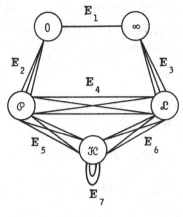

Fig. 9

Proof. We note that $|X_1| = 1 + |\mathcal{P}| = 22$, and $|\mathcal{B}_1| = |\mathcal{H}| + |\mathcal{L}| = 56 + 21 = 77$, so that we have the correct number of points and blocks. We have to show that 'k = 6' and '$\lambda = 1$', in the usual notation for designs.

<u>k = 6</u> Suppose $l \in \mathcal{L}$. Then l is adjacent in Σ to ∞ and to the five points $p \in \mathcal{P}$ which lie on it in the plane $(\mathcal{P}, \mathcal{L})$. Hence there are six members of X_1 I_1-incident with l.

Suppose $H \in \mathcal{H}$; then the six points of $\mathcal{P}$ belonging to H are adjacent to H in Σ and consequently I_1-incident with H.

<u>$\lambda = 1$</u> Consider a 3-subset of X_1 of the form $\{\infty, p_1, p_2\}$ $(p_1, p_2 \in \mathcal{P})$. The points p_1 and p_2 determine a unique line l, and ∞, p_1, p_2 are I_1-incident with l.

For a 3-subset of X_1 consisting of three points p_1, p_2, p_3, there are two possibilities. First, if p_1, p_2, p_3 are collinear on a line l, then they are I_1-incident with l. Secondly, suppose that p_1, p_2, p_3 are not collinear: we shall show that there is a unique hyperoval in $\mathcal{H}$ which contains these points. There are 168 hyperovals altogether (3.9.3) and each one contains $\binom{6}{3} = 20$ sets of three non-collinear points. There are $(21.20.16)/(3.2.1) = 1120$ sets of three non-collinear points altogether, and so each one is contained in μ hyperovals, where $168.20 = 1120\mu$. Thus $\mu = 3$. The three hyperovals containing $\{p_1, p_2, p_3\}$ intersect pairwise in these three points, and so no two of them can lie in the same 'extension class' (3.9.2). Hence precisely one of them belongs to the chosen class $\mathcal{H}$, and p_1, p_2, p_3 are I_1-incident with it.

This completes the proof that $(X_1, \mathcal{B}_1, I_1)$ is a $3 - (22, 6, 1)$ design. $\parallel$

4.5.2 Lemma. <u>Let</u> $X_2 = \{0\} \cup \mathcal{L}$, $\mathcal{B}_2 = \mathcal{P} \cup \mathcal{H}$, <u>and define an incidence relation</u> I_2 <u>by the rule that</u>

$$x I_2 \beta \Longleftrightarrow \{x, \beta\} \text{ is an edge of } \Sigma .$$

<u>Then</u> $(X_2, \mathcal{B}_2, I_2)$ <u>is a</u> $3 - (22, 6, 1)$ <u>design.</u>

Proof. The arguments are very similar to those in the previous proof, and we shall use the same framework.

<u>k = 6</u> Suppose $p \in \mathcal{P}$. Then p is adjacent in Σ to 0 and to the five lines l which contain it in the plane $(\mathcal{P}, \mathcal{L})$. Hence there are six members of X_2 which are I_2-incident with p.

Suppose $H \in \mathcal{H}$; the points of H determine $\binom{6}{2} = 15$ lines, and so there are just $6 = 21 - 15$ lines not meeting H. These are the six members of X_2 which are I_2-incident with H.

<u>$\lambda = 1$</u> Consider a 3-subset of X_2 of the form $\{0, l_1, l_2\}$ $(l_1, l_2 \in \mathcal{L})$. The lines l_1 and l_2 determine a unique point p, and $0, l_1, l_2$ are I_2-incident with it.

For a 3-subset of X_2 consisting of three lines l_1, l_2, l_3, there are two possibilities. First, if l_1, l_2, l_3 are concurrent in a point p, then they are I_2-incident with p. (The lines cannot all be I_2-incident with a hyperoval, since (by 2.9.4) at most two of the lines through a point can be disjoint from a hyperoval.) Secondly, suppose that l_1, l_2, and l_3 are not concurrent: we shall show that there is a unique hyperoval in $\mathcal{H}$ which is disjoint from these lines. There are 168 hyperovals altogether, and each one is disjoint from $\binom{6}{3} = 20$ sets of three non-concurrent lines. There are $(21.20.16)/(3.2.1) = 1120$ sets of three non-concurrent lines altogether, and so each one is disjoint from 3 hyperovals, by the same counting argument as before. The three lines contain 12 points, and so the three hyperovals disjoint from them are subsets of the remaining 9 points. Thus any pair of the three hyperovals must intersect

96

in at least three points, and precisely one of them belongs to the chosen class $\mathcal{K}$. This is the unique element of $\mathcal{B}_2$ which is I_2-incident with the three given lines. //

We now recall that there is a unique 3 - (22, 6, 1) design, and that M_{22} is a group of automorphisms of it, acting transitively on points and blocks. The two designs imbedded in the graph Σ force the existence of two groups isomorphic to M_{22} acting as graph isomorphisms of Σ.

4.5.3 **Lemma.** The stabilizers of the vertices 0 and ∞ in the automorphism group of Σ both contain subgroups isomorphic to M_{22}.

Proof. Let g be any element of M_{22}, acting as an automorphism of the design $(X_1, \mathcal{B}_1, I_1)$; in other words, g is a permutation of $X_1 = \mathcal{P} \cup \{\infty\}$ and a permutation of $\mathcal{B}_1 = \mathcal{K} \cup \mathcal{L}$ such that $x I_1 \beta$ if and only if $gx I_2 g\beta$. Define $g(0) = 0$, so that g is a permutation of the entire vertex-set of Σ. Since g fixes 0 it preserves the edges in $E_1 \cup E_2$, and the definition of I_1 ensures that g preserves $E_3 \cup E_4 \cup E_5$. Finally, E_6 and E_7 are preserved, since g must take disjoint blocks to disjoint blocks. Thus g is an automorphism of the graph Σ.

Similarly, any element of M_{22}, acting as an automorphism of the design $(X_2, \mathcal{B}_2, I_2)$, induces a graph automorphism of Σ fixing the vertex ∞. //

4.5.4 **Definition.** The Higman-Sims group H is the subgroup of Aut Σ generated by the two subgroups M_{22} in Lemma 4.5.3.

4.5.5 **Theorem.** The group H acts transitively, with rank 3, on the vertices of Σ, and the parameters of Σ as a strongly regular graph are (22, 0, 6).

Proof. Let us denote the two copies of M_{22} by $M_{22}^{(0)}$ and $M_{22}^{(\infty)}$, with the obvious meaning. Suppose that $p \in \mathcal{P}$, $l \in \mathcal{L}$, and $K \in \mathcal{K}$ are given. Since $M_{22}^{(0)}$ is transitive on the points and blocks of the I_1-design, we can find automorphisms g_0 and h_0 of Σ, fixing 0, such that

$g_0(\infty) = p$ and $h_0(K) = l$. Similarly, we can find automorphisms g_∞ and h_∞, fixing ∞, such that $g_\infty(0) = l$ and $h_\infty(K) = p$. Suitable combinations of these four automorphisms and their inverses take 0 to ∞, p, l, and K, and so the group H (generated by $M_{22}^{(0)}$ and $M_{22}^{(\infty)}$) is transitive on the vertices of Σ.

H_0, the stabilizer of 0, has three orbits $\{0\}$, $\mathcal{P} \cup \{\infty\}$, and $\mathcal{H} \cup \mathcal{L}$, so that H is a rank 3 group. According to 4.3.1, Σ is a strongly regular graph; its parameters $(22, 0, 6)$ may be found by taking the initial vertex to be 0 and comparing Fig. 8 with the construction of Σ. //

It remains only to show that H is a simple group. For this, we return to the fundamental results proved in Chapter 1. In order to apply Theorem 1.6.7 we have to make the remark that the stabilizer H_0 is just $M_{22}^{(0)}$: by the symmetry of the construction we must have

$$(M_{22}^{(0)})_\infty = (M_{22}^{(\infty)})_0,$$

and so the elements of $M_{22}^{(\infty)}$ fixing 0 are already in $M_{22}^{(0)}$, and $H_0 = M_{22}^{(0)}$. Thus H_0 is a simple group.

4.5.6 Theorem. <u>The Higman-Sims group H is a simple group.</u>

Proof. The stabilizer H_0 is the simple group M_{22}. Also, the parameters $\kappa = 22$ and $\gamma = 6$ of Σ are not equal, and so (by 4.3.4) H acts primitively on the vertices of Σ. Thus, in view of the result 1.6.7, we have only to prove that H has no regular normal subgroups.

Suppose that N is a regular normal subgroup of H; then $|N| = 100 = 2^2 \cdot 5^2$, since this is the number of vertices of Σ. The number s of Sylow 5-subgroups of N is a divisor of 100 and satisfies $s \equiv 1 \pmod 5$ (see Section 1.8). Hence $s = 1$, and N has a unique Sylow 5-subgroup P; P is a characteristic subgroup of N, and hence it is a normal subgroup of H. But since H acts primitively on the vertices, P must act transitively, which is impossible since $|P| = 25$. This contradiction shows that H has no regular normal subgroups and so H is simple. //

The order of H is $100. |M_{22}| = 44,352,000$. The fact that it is a sporadic simple group can be established by comparing its order with the orders of the known simple groups belonging to infinite families.

4.6 Project: Some graphs and their automorphism groups

If we are given a graph Γ, it is sometimes easy to prove that a certain group G acts as a group of automorphisms of Γ, but difficult to check that G is the full automorphism group $\text{Aut}\,\Gamma$. One possible approach is to use the structure of the graph to find a bound for $|\text{Aut}\,\Gamma|$. Of course, when Γ is known to be vertex-transitive it is sufficient to find a bound for $|(\text{Aut}\,\Gamma)_x|$, $x \in V$.

Suppose that $\Gamma = (V, E)$ and x, y are vertices such that $\{x, y\}$ is an edge. Let G be a group of automorphisms of Γ, acting transitively on the vertex set V and on the set of ordered pairs of adjacent vertices. Define

$$L_x = \{g \in G_x | g \text{ fixes each vertex adjacent to } x\};$$

$$L_{xy} = L_x \cap L_y.$$

4.6.1 Let κ denote the valency of Γ. Show that there is a homomorphism of G_x into a permutation group of degree κ, with kernel L_x, and deduce that $|G_x : L_x| \leq \kappa!$.

4.6.2 Verify the following statements (including the implicit results concerning the normality of subgroups):

$$\frac{L_x}{L_{xy}} \approx \frac{L_x L_y}{L_y} \trianglelefteq \frac{G_{xy}}{L_y} \; .$$

Show that $|L_x : L_{xy}| \leq (\kappa - 1)!$, and deduce that if $L_{xy} = 1$, then $|G_x| \leq \kappa! (\kappa - 1)!$.

4.6.3 The complete bipartite graph $K_{n,n}$ is defined as follows. $V = A \cup B$, where A and B are disjoint n-sets, and E is the set of all pairs $\{a, b\}$ $(a \in A, b \in B)$. Use 4.6.2 to show that

$$|\text{Aut}\,K_{n,n}| \leq 2(n!)^2,$$

and hence describe the full automorphism group of $K_{n, n}$.

4.6.4   The <u>odd graph</u> O_κ is defined as follows (for $\kappa \geq 2$). V is the set of $(\kappa-1)$-subsets of a fixed $(2\kappa-1)$-set, and E is the set of pairs $\{v, w\}$ such that v and w are disjoint sets. (For example, O_2 is a triangle, and O_3 is Petersen's graph.) Show that the symmetric group $S_{2\kappa-1}$ acts as a group of automorphisms of O_κ, and that it is transitive on vertices and on ordered pairs of adjacent vertices. Find its rank as a permutation group on the vertices.

4.6.5   Let a, b, c, d be four distinct vertices of O_κ, such that $\{a, b\}$, $\{b, c\}$, $\{c, d\}$ are edges. Show that a, b, c, d determine a unique pair of vertices $e \neq c$ and $f \neq b$ such that $\{d, e\}$, $\{e, f\}$ and $\{f, a\}$ are edges.

4.6.6   Let G be the full group of automorphisms of O_κ. Use 4.6.5 to show that if $\alpha \in L_{ab}$, then $\alpha \in L_{bc}$. Deduce that $L_{xy} = 1$ for all adjacent pairs (x, y).

4.6.7   Show that $G = S_{2\kappa-1}$.

4.7   Project: Strongly regular graphs and biplanes

We shall investigate strongly regular graphs which have no triangles - that is, no triples of mutually adjacent vertices. In terms of the parameters (κ, α, γ) this means that $\alpha = 0$. We shall see that the feasibility conditions impose severe restrictions on κ and γ, and that in the case $\gamma = 2$ there is a relationship with an interesting class of symmetric designs.

4.7.1   Use 4.4.4 to show that if the parameters $(\kappa, 0, \gamma)$, $\gamma \geq 2$, are feasible then $\gamma^2 - 4\gamma + 4\kappa$ is the square of an integer s, and

$$m = \frac{\kappa}{2\gamma s}((\kappa - 1 + \gamma)(s + \gamma) - 2\gamma)$$

is also an integer. Eliminate κ and express the resulting identity as a polynomial equation in s. Deduce that for each value of $\gamma \neq 2$, 4, 6 there are only finitely many corresponding values of κ.

4.7.2 Show that the parameters $(\kappa, 0, 2)$ are feasible if and only if $\kappa = t^2 + 1$, where t is an integer not congruent to 0 (mod 4).

4.7.3 Let Γ be a strongly regular graph with parameters $(\kappa, 0, 2)$, and let X be its vertex-set. Define

$$\beta_x = \{y \in X | y = x \text{ or } y \text{ is adjacent to } x\},$$
$$\mathcal{B} = \{\beta_x | x \in X\}.$$

Show that $(X, \mathcal{B})$ is a symmetric $2 - (\frac{1}{2}(\kappa^2 + \kappa + 2), \kappa + 1, 2)$ design. (Such a design is called a biplane.) By 4.7.2, biplanes could exist for $k = t^2 + 1$, $t = 1, 2, 3, 5, \dots$. Find the graph and the corresponding biplane for $t = 1$.

4.7.4 Suppose Γ_5 is a strongly regular graph with parameters $(5, 0, 2)$. Let $*$ be any vertex and label the adjacent vertices $1, 2, 3, 4, 5$. Show that the 10 vertices not adjacent to $*$ must be labelled $12, 13, \dots, 45$ in a canonical way, and that all the edges of the graph are determined uniquely. Show that Γ_5 admits a transitive, rank 3, group of automorphisms, so that we have in fact constructed a strongly regular graph.

4.7.5 Define a graph Γ_{10} as follows. The vertices are the 56 hyperovals in $PG(2, 4)$ which belong to a given 'extension class', and the edges join disjoint hyperovals. Prove that Γ_{10} is a strongly regular graph with parameters $(10, 0, 2)$. (Hint: Γ_{10} is a subgraph of the Higman-Sims graph.)

NOTES AND REFERENCES FOR CHAPTER 4

Basic references for the graphical representation of permutation groups are the papers of Higman [4], [5], and Sims [7]. Strongly regular graphs are a special case of 'distance-regular' graphs, and the linear algebra of Section 4.4 may be extended to the general case [1].

The Higman-Sims group was discovered in 1968 [6]. Since that time several other sporadic simple groups have been constructed as rank 3 groups of automorphisms of graphs. A survey may be found in [3].

Many interesting results concerning the relationship between strongly regular graphs and designs are given in the book of Cameron and van Lint [2].

1.	Biggs, N. L. Algebraic Graph Theory. (Cambridge University Press, 1974.)

2.	Cameron, P. J. and Lint, J. H. van. Graph Theory, Coding Theory and Block Designs. (Cambridge University Press, 1975.)

3.	Feit, W. The current situation in the theory of finite simple groups. Proceedings of the ICM (1970), 1, 55-93.

4.	Higman, D. G. Finite permutation groups of rank 3. Math. Zeitschr., 86 (1964), 145-56.

5.	Higman, D. G. Intersection matrices for finite permutation groups. J. Algebra, 6 (1967), 22-44.

6.	Higman, D. G. and Sims, C. C. A simple group of order 44,352,000. Math. Zeitschr., 105 (1968), 110-13.

7.	Sims, C. C. Graphs and finite permutation groups. Math. Zeitschr. 95 (1967), 76-86.

5 · Maps

'The Descriptive-Geometry Theorem that any map whatever can have its
divisions properly distinguished by the use of but four colours, from its
generality and intangibility, seems to have aroused a good deal of interest
some few years ago ...'

P. J. Heawood, in his paper Map-Colour Theorem, 1890.

5.1 Maps and surfaces

In this chapter we shall discuss graphs realized by a set of points and
lines on a closed orientable surface. Although this notion arises in a
topological context, we shall show that it is possible to develop it by
purely combinatorial arguments. The reader who is unfamiliar with the
topology of surfaces should not be at a disadvantage.

We begin at an intuitive level. For our purposes, it is sufficient
to say that a 'surface' is a compact topological space which has two
special properties:

(i)   it is locally homeomorphic to ordinary Euclidean 2-space;

(ii)   it has a consistent global orientation.

The sphere and the torus are the simplest examples. The Euclidean plane
is not compact, and so it is not a 'surface' for us; however, it may be
made homeomorphic to a sphere by the addition of a single point, and so
the two spaces have very similar properties. The Klein bottle is not
allowed, since it has no consistent global orientation.

Let us suppose that a graph Γ is represented by a set of points
and lines on a surface, in such a way that the lines intersect only at the
points representing their end vertices. The lines divide the surface into
connected regions, called 'faces', and the resulting configuration is a
'map'. (We shall require, for technical reasons, that each face is homeomorphic to an open disc.) Thus a 'map' has points (vertices), lines
(edges), and regions (faces). At each vertex v, the neighbourhood of v

is locally like a plane, and so the vertices adjacent to v have a cyclic ordering ρ_v corresponding to the arrangement of the edges joining them to v on the surface (Fig. 10).

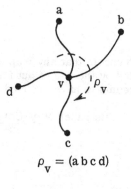

$$\rho_v = (a\ b\ c\ d)$$

Fig. 10

It turns out that the family of cyclic permutations $\{\rho_v\}$, one for each vertex v of Γ, is an adequate combinatorial description for the topological 'map'. We give no more justification for this statement than is provided by the foregoing remarks. Thus, we do not need a definition of 'surface', and our definition of 'map' will be purely combinatorial. The remainder of this chapter is a consistent and self-contained piece of mathematics, although we shall refer to the intuitive concepts for motivation and illustration.

5.1.1 Definition. A <u>rotation</u> on a graph $\Gamma = (V,\ E)$ is a set

$$\rho = \{\rho_v\}_{v \in V},$$

where each ρ_v is a cyclic permutation of the vertices adjacent to v in Γ. A <u>map</u> is a pair $(\Gamma,\ \rho)$, where Γ is a connected graph and ρ is a rotation on Γ.

For example, a rotation on the complete graph K_4 is tabulated on the left below (Fig. 11). On the right, there is a drawing of K_4 which may be regarded as the map corresponding to the given rotation. (The plane of the paper is to be thought of as a portion of a sphere.)

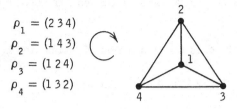

$$\rho_1 = (2\,3\,4)$$
$$\rho_2 = (1\,4\,3)$$
$$\rho_3 = (1\,2\,4)$$
$$\rho_4 = (1\,3\,2)$$

Fig. 11

We turn to the formal definition of the faces of a map. Let $S = S\Gamma$ denote the set of <u>sides</u> of Γ:

$$S\Gamma = \{(v,\,w)\,|\,\{v,\,w\}\text{ is an edge of }\Gamma\}.$$

Thus each edge $\{v,\,w\}$ gives rise to two sides $(v,\,w)$ and $(w,\,v)$. We often think of $(v,\,w)$ as a directed line pointing from v to w. A rotation ρ on Γ induces a permutation of S (also denoted by ρ):

$$\rho(v,\,w) = (v,\,\rho_v(w)).$$

This corresponds to rotating the sides pointing away from v in the order prescribed by ρ_v. The cycles of ρ on S are in one-to-one correspondence with the vertices of Γ. Now we define a permutation ρ^* of S, as follows:

$$\rho^*(v,\,w) = (w,\,\rho_w(v)) = \rho(w,\,v).$$

Fig. 12 illustrates ρ and ρ^* acting on S.

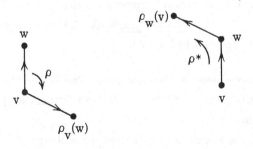

Fig. 12

5.1.2 Definition. Let $M = (\Gamma, \rho)$ be a map, and suppose ρ^* is as defined above. A <u>face</u> of M is a cyclic sequence of vertices occurring in a cycle of ρ^* on S.

For example, if ρ is the rotation on K_4 tabulated in Fig. 11, then

$$\rho^* = (12\ 24\ 41)(13\ 32\ 21)(14\ 43\ 31)(23\ 34\ 42),$$

and the faces are $(1, 2, 4)$, $(1, 3, 2)$, $(1, 4, 3)$, and $(2, 3, 4)$. It may be checked that these correspond to the regions of the figure, including the 'outside' region. As another example, suppose we define a rotation on K_7 as follows:

$$\rho_1 = (243756)$$
$$\rho_2 = (354167)$$
$$\rho_3 = (465271)$$
$$\rho_4 = (576312)$$
$$\rho_5 = (617423)$$
$$\rho_6 = (721534)$$
$$\rho_7 = (132645).$$

There are fourteen faces, $(1, 2, 6)$, ... In this case we cannot represent the map on a sphere or a plane without some further conventions. (See Fig. 20.)

One of the oldest results in the theory of maps concerns a relationship linking the numbers of vertices, edges, and faces. In order to obtain this result by combinatorial means we need some simple lemmas.

5.1.3 Lemma. <u>Given any connected graph</u> $\Gamma = (V, E)$ <u>we can find a subset</u> T <u>of</u> E <u>such that the graph</u> (V, T) <u>is connected and has no circuits. Furthermore,</u> $|T| = |V| - 1$.

Proof. We use induction on $n = |V|$. The result is true when $n = 2$, since the only connected graph with two vertices has just one edge e, and $T = \{e\}$ is the required subset.

Suppose that the result is true for graphs with at most $n - 1$ vertices, and let Γ be a connected graph with $|V| = n$. Choose v in V and let

106

E' denote the set of edges not incident with v. The graph $\Gamma' = (V - \{v\},\ E')$ is the union of disjoint connected graphs $\Gamma_\lambda = (V_\lambda,\ E_\lambda)$, and by the induction hypothesis, each E_λ contains a subset T_λ such that $(V_\lambda,\ T_\lambda)$ is connected and has no circuits. Also, since Γ is connected, there is an edge e_λ joining v to some vertex v_λ in V_λ. Let T be the union of all the edges e_λ and all the edges in the sets T_λ. We have

$$|T| = \sum_\lambda |T_\lambda \cup \{e_\lambda\}|$$

$$= \sum_\lambda |V_\lambda| \qquad \text{(by the induction hypothesis)}$$

$$= |V - \{v\}| = |V| - 1.$$

Thus T is the required subset of E. $/\!/$

The set T, or the graph (V, T), is called a <u>spanning tree</u> for Γ.

If π is any permutation of a finite set X, and $\sigma = (xy)$ is a transposition, then it is easy to check that $c(\pi)$, the number of cycles of π, is related to $c(\pi\sigma)$ as follows:

$$5.1.4 \quad c(\pi\sigma) = \begin{cases} c(\pi) + 1 & \text{if } x \text{ and } y \text{ are in the same cycle of } \pi; \\ c(\pi) - 1 & \text{if not.} \end{cases}$$

We shall require this result applied to the set S of sides of a graph $\Gamma = (V,\ E)$. Suppose that $D \subseteq E$, and τ_D is the permutation of S defined by

$$\tau_D(v,\ w) = \begin{cases} (w,\ v) & \text{if } \{v,\ w\} \in D; \\ (v,\ w) & \text{if not.} \end{cases}$$

Thus τ_D is the composition of transpositions τ_e ($e \in D$); τ_e switches the two sides corresponding to the edge e and fixes every other side.

5.1.5 **Lemma.** <u>If $(\Gamma,\ \rho)$ is a map and T is a spanning tree for Γ, then $\rho\tau_T$ has just one cycle on</u> SΓ.

Proof. Let $T = \{e_1, \ldots, e_{n-1}\}$ where n is the number of vertices of Γ. By its definition, τ_T is the composition of $n-1$ transpositions τ_i, such that τ_i switches the sides corresponding to e_i. Thus $\rho\tau_T = \rho\tau_1\tau_2 \cdots \tau_{n-1}$.

Since the two sides of e_1 are in different cycles of ρ acting on $S = S\Gamma$, we have $c(\rho\tau_1) = c(\rho) - 1 = n - 1$. Now we may proceed inductively: suppose that $c(\rho\tau_1 \cdots \tau_m) = n - m$, for some value of m $(1 \leq m < n-1)$; then the equation 5.1.4 tells us that

$$c(\rho\tau_1 \cdots \tau_{m+1}) = n - m \pm 1.$$

If the $+$ sign holds, the two sides of e_{m+1} must be in the same cycle of $\rho\tau_1 \cdots \tau_m$ on S; this means that there is a chain of edges selected from $\{e_1, \ldots, e_m\}$ joining the two end vertices of e_{m+1} - in other words, we have a circuit in T. Thus the minus sign must hold, and the induction step is complete. It follows that

$$c(\rho\tau_T) = c(\rho\tau_1 \cdots \tau_{n-1}) = n - (n - 1) = 1. \;\;/\!/$$

5.1.6 Theorem. <u>Let $\Gamma = (V, E)$ be a connected graph and ρ a rotation on Γ. Let F denote the set of faces of the map $M = (\Gamma, \rho)$. There is a non-negative integer $g(M)$ such that</u>

$$|V| - |E| + |F| = 2 - 2g(M).$$

Proof. Let T be a spanning tree for Γ, and let $U = E - T$. We remark that ρ^* is just $\rho\tau_E$, since

$$\rho\tau_E(v, w) = \rho(w, v) = (w, \rho_w(v)) = \rho^*(v, w).$$

Now $|F|$ is, by definition, the number of cycles of ρ^*, and so

$$|F| = c(\rho^*) = c(\rho\tau_E) = c(\rho\tau_T\tau_U).$$

The lemma has established that $c(\rho\tau_T) = 1$. To obtain ρ^* from $\rho\tau_T$ we form the composition with the transpositions ρ_e ($e \in U$). As each transposition is added, the number of cycles either increases or decreases by unity. Suppose that it increases h times and decreases g

times; we have

$$h + g = |U| = |E - T| = |E| - |V| + 1,$$
$$1 + h - g = c(\rho^*) = |F|.$$

Eliminating h gives

$$|V| - |E| + |F| = 2 - 2g,$$

as required. //

5. 1. 7 **Definition.** The non-negative integer g(M) occurring in Theorem 5. 1. 6 is called the genus of the map M.

We have defined the genus of a map in a purely combinatorial way. Readers who are familiar with the topological background will see that a map (Γ, ρ) of genus g is an imbedding of Γ in the surface of a sphere with g handles, all regions being open discs. In particular, when g = 0, Γ is represented on the sphere itself, and thence, by omitting a suitable point, on the plane. Thus we say that a graph Γ, for which there exists some ρ such that the genus of (Γ, ρ) is zero, is a planar graph. Of course, the genus g depends on ρ as well as Γ, so that, for example, K_4 has rotations which give rise to maps of genus 0 and 1.

5. 2 Automorphisms of maps

We shall define an automorphism of a map (Γ, ρ) to be an automorphism of the graph Γ which also 'preserves' the rotation ρ. In order to make this idea precise, we begin by constructing an action of Aut Γ on the set $R(\Gamma)$ of all rotations of Γ. If $g \in$ Aut Γ and $\rho \in R(\Gamma)$, we define $\rho^{(g)}$ in $R(\Gamma)$ as follows:

$$\rho^{(g)}_{gv} = g\rho_v g^{-1}.$$

In other words, the rotation $\rho^{(g)}$ at the vertex gv takes gx to gy when ρ_v takes x to y (Fig. 13).

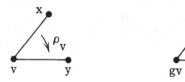

Fig. 13

5.2.1 Definition. Two rotations ρ and σ in $R(\Gamma)$ are equivalent if $\sigma = \rho^{(g)}$ for some automorphism g of Γ.

5.2.2 Lemma. Suppose that ρ and σ are equivalent rotations on Γ, so that $\sigma = \rho^{(g)}$, and let $(x, y, z \ldots t)$ be a face of the map (Γ, ρ). Then $(gx, gy, gz \ldots gt)$ is a face of the map (Γ, σ).

Proof. Since $(x, y, z \ldots t)$ is a face of (Γ, ρ), we have $\rho^*(x, y) = (y, z)$, that is, $\rho_y(x) = z$. Hence, by definition of $\rho^{(g)}$,

$$\rho_{gy}^{(g)}(gx) = g\rho_y g^{-1}(gx) = g\rho_y(x) = gz.$$

Thus, putting $\sigma = \rho^{(g)}$, we have $\sigma_{gy}(gx) = gz$ and

$$\sigma^*(gx, gy) = (gy, gz).$$

This shows that $(gx, gy, gz \ldots gt)$ is a face of (Γ, σ). //

The lemma shows that, when ρ and σ are equivalent rotations, there is a one-to-one correspondence between the faces of the maps (Γ, ρ) and (Γ, σ); in particular, the two maps have the same genus.

5.2.3 Definition. An automorphism of a map $M = (\Gamma, \rho)$ is an automorphism g of the graph Γ such that $\rho^{(g)} = \rho$; that is, $\rho_{gv} = g\rho_v g^{-1}$ for each v in $V\Gamma$.

Thus an automorphism of (Γ, ρ) is an element of Aut Γ which fixes ρ in the action of Aut Γ on $R(\Gamma)$. Lemma 5.2.2 tells us that a map automorphism permutes the faces of the map in the obvious way: if $(x, y, z \ldots t)$ is a face of M and g is an automorphism of M, then $(gx, gy, gz \ldots gt)$ is also a face of M.

We shall write Aut M, or Aut(Γ, ρ), for the group formed by all automorphisms of the map M = (Γ, ρ).

5.2.4 Theorem. <u>Let Γ be a connected graph and ρ a rotation on Γ.</u>

(i)   <u>The number of rotations equivalent to ρ is equal to the index</u> $\left| \text{Aut } \Gamma : \text{Aut}(\Gamma, \rho) \right|$.

(ii)   <u>The number of equivalence classes of maps with underlying graph Γ is</u>

$$\left| \text{Aut } \Gamma \right|^{-1} \sum_{g \in \text{Aut } \Gamma} \left| \Phi(g) \right| ,$$

<u>where $\Phi(g)$ is the number of rotations ρ on Γ such that $\rho^{(g)} = \rho$.</u>

Proof.   (i) As we have remarked, Aut(Γ, ρ) is the stabilizer of ρ in the action of Aut Γ on R(Γ). Hence the basic result 1.2.4 gives what is needed.

(ii)   This is simply Burnside's Lemma (1.2.5), applied to the action of Aut Γ on R(Γ). //

We may illustrate these ideas by referring to the complete graph K_4. Since there are two possible cyclic permutations of the three vertices adjacent to a given vertex, there are $2^4 = 16$ rotations in all. One of them is shown in Fig. 11, and it may be checked that the graph automorphism (12)(34) is a map automorphism in this case, whereas, for example, the graph automorphism (12) is not a map automorphism. In fact, the map automorphism group is the subgroup A_4 of the graph automorphism group S_4. Since $\left| S_4 : A_4 \right| = 2$, we deduce from 5.2.4(i) that there are just two rotations equivalent to the given one. In order to classify all 16 rotations of K_4, we note that $\Phi(g)$ is a class function - that is, it is constant on each conjugacy class of elements of S_4. We easily obtain the following table:

Class representative (g)	1	(12)	(123)	(12)(34)	(1234)
Number in class	1	6	8	3	6
$\left| \Phi(g) \right|$	16	0	4	4	2 .

Thus the number of equivalence classes is $(16 + 32 + 12 + 12)/24 = 3$.
One class is represented by the rotation in Fig. 11, and we have seen that
this class contains just two rotations. The other two classes both give
rise to maps with genus 1, and to illustrate them diagrammatically we
need a conventional representation of the 'surface' of genus 1, or
'torus'. This is obtained by identifying the pairs of opposite sides of a
plane rectangle. Representatives of the two toroidal classes are shown
in Fig. 14. It is worth noting that these maps have faces with 'repeated'
vertices: the faces of the map (a) are (0, 3, 1, 2) and (0, 1, 3, 2, 1,
0, 2, 3).

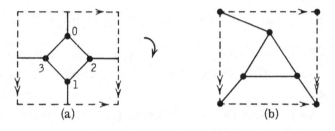

(a)   (b)

Fig. 14

In general, the number of rotations of a graph is very large, and
the number of equivalence classes under the action of the automorphism
group is also large. For example, the numbers of rotations for K_5 and
K_6 respectively are $7,776$ and $191,102,976$, while the numbers of
classes are 78 and $265,764$.

We now turn to the general structure of $\text{Aut}(\Gamma, \rho)$, regarded as
a group of permutations of the vertex-set of Γ. The following simple
lemma is fundamental.

5.2.5 Lemma. Let $M = (\Gamma, \rho)$, $g \in \text{Aut } M$, and suppose that
$\{v, w\}$ is an edge of Γ. If g fixes both v and w, then g is the
identity automorphism.

Proof. Using the facts that $g(v) = v$, $g(w) = w$, and $\rho^{(g)} = \rho$,
we may compute as follows:

$$g[\rho_w(v)] = g\rho_w g^{-1}(v) = \rho_{gw}^{(g)}(v) = \rho_w(v).$$

Thus g fixes the vertex $\rho_w(v)$ also. Since ρ_w is cyclic on the set of vertices adjacent to w, we may repeat the argument and deduce that g fixes all vertices adjacent to w. Similarly, g fixes all vertices adjacent to v, and all vertices adjacent to $\rho_w(v)$, ... Finally, since Γ is connected, we may conclude that g fixes all vertices of Γ, and so g is the identity. //

Let g be an element of Aut M fixing the vertex v. Since g is a graph automorphism, it permutes the set $N(v)$ of vertices adjacent to v, and we have a restricted permutation $\bar{g} = g \,|\, N(v)$. If h is also an automorphism of M fixing v, and $\bar{g} = \bar{h}$, then hg^{-1} fixes two adjacent vertices and so $g = h$. Thus the stabilizer of v in Aut M is determined by its action on $N(v)$. In fact, we can be still more precise about the stabilizer.

5.2.6 Theorem. <u>Let</u> $A = $ Aut M, <u>where</u> M <u>is the map</u> (Γ, ρ), and let v be a vertex of Γ. The stabilizer A_v is isomorphic to a subgroup of the cyclic group $\langle \rho_v \rangle$ generated by ρ_v. Thus A_v is a cyclic group and its order divides the valency of v.

Proof. If g is in A_v, then

$$g\rho_v g^{-1} = \rho_{gv} = \rho_v ,$$

since $\rho^{(g)} = \rho$ and $gv = v$. Thus $g\rho_v = \rho_v g$.

Let w be a vertex adjacent to v; $g(w)$ is also adjacent to v, and since ρ_v is cyclic on $N(v)$ we have $g(w) = \rho_v^i(w)$, for some i. Let x be any vertex adjacent to v, say $x = \rho_v^j(w)$. Then

$$g(x) = g\rho_v^j(w) = \rho_v^j g(w) = \rho_v^{j+i}(w) = \rho_v^i \rho_v^j(w) = \rho_v^i(x).$$

Thus $\bar{g}$, the restriction of g to $N(v)$, is equal to ρ_v^i. It follows that $g \mapsto \bar{g}$ is a mapping of A_v into $\langle \rho_v \rangle$, and the remarks preceding the theorem show that it is a monomorphism. Hence we have the result. //

If $M = (\Gamma, \rho)$ and $A = $ Aut M is transitive on the vertices of Γ, then we say that M is a <u>vertex-transitive map.</u> The general theory developed in Chapter 1 shows that all the vertex-stabilizers A_v are conjugate in A, and $|A| = n\ |A_v|$, where n is the number of vertices of Γ. Theorem 5.2.6 tells us that $|A| = n\delta$, where δ is a divisor of κ, the valency of the graph.

An example of a vertex-transitive map is provided by the rotation on K_7 given in Section 5.1. It is easy to see that the permutation $g = (1234567)$ is an automorphism of this map, so that the automorphism group A is transitive on the vertices. Furthermore, $h = (243756)$ is an automorphism fixing 1 (note that $h = \rho_1$), and so $|A| = 7.6 = 42$. Of the three classes of maps involving K_4, two are vertex-transitive and one is not.

We have seen that the order of the group of a vertex-transitive map is $n\delta$, where δ divides the valency κ. Thus the greatest symmetry occurs when $\delta = \kappa$. In that case $|A| = n\kappa$, and this number is equal to the number of sides, or twice the number of edges, of Γ.

5.2.7 Definition. A map $M = (\Gamma, \rho)$ is <u>symmetrical</u> if it is vertex-transitive and $|\text{Aut } M| = 2|E\Gamma|$.

An equivalent condition is that $\text{Aut } M$ is transitive on the vertices and ρ_v is the restriction to $N(v)$ of an automorphism of M. Yet another formulation arises when we consider the action of $\text{Aut } M$ on the set S of sides of Γ. If $(v, w) \in S$ and $g \in \text{Aut } M$, then $(gv, gw) \in S$, since g is a graph automorphism. This action of $\text{Aut } M$ on S cannot be more than sharply 1-transitive (that is, regular), since $|\text{Aut } M| \leq |S|$; it is regular precisely when $|\text{Aut } M| = |S|$. In other words, $\text{Aut } M$ acts regularly on S if and only if M is a symmetrical map. The term 'regular map' is often used as a synonym for a symmetrical map, and the preceding observation indicates that the term is an apt one; however, it is unlikely that it was introduced with this relationship in view.

Given a graph Γ, we may ask whether there is a rotation ρ such that the map (Γ, ρ) is symmetrical. Clearly, Γ itself must have enough automorphisms ($\text{Aut } \Gamma$ must be transitive on the sides of Γ), but this is not sufficient. For example, the Petersen graph has enough automorphisms, since the work in Section 4.6 establishes that its group is S_5. However, a symmetrical map M for this graph would have

$|\text{Aut } M| = 2|E| = 30$, and S_5 has no subgroup of order 30.

In Section 5.4 we shall investigate the possibility of symmetrical maps when the underlying graph is a complete graph.

5.2.8 Theorem. If M is a symmetrical map, then Aut M acts transitively on the vertices, edges, and faces of M.

Proof. Aut M is transitive on the vertices, by definition. Also, we have remarked that it is transitive on the set of sides, and hence it is transitive on the edges. Suppose that $f_1 = (v, w, \ldots)$ and $f_2 = (x, y, \ldots)$ are two faces of M. There is an automorphism g of M taking the side (v, w) to (x, y), and since

$$g\rho_w g^{-1} = \rho_{gw} = \rho_y,$$

g also takes the next vertex $\rho_w(v)$ of f_1 to the next vertex $\rho_y(x)$ of f_2. Clearly, we can repeat this argument, and show that g takes the vertices of f_1 to those of f_2 in order. Thus Aut M is transitive on the faces of M. //

It follows that there are two important constants associated with a symmetrical map: the constant valency (κ) of each vertex, and the constant number (l) of vertices in each face. In Section 5.5 we shall investigate the relationship between these constants κ, l and the genus g.

5.3 Cayley graphs and Cayley maps

The action of the group Aut M on the vertex-set of a map $M = (\Gamma, \rho)$ is most conveniently described when Γ is a Cayley graph.

5.3.1 Definition. Let G be a group, with $\Omega \subseteq G$ such that:
(i) Ω generates G;
(ii) $1 \notin \Omega$;
(iii) $\omega \in \Omega \Rightarrow \omega^{-1} \in \Omega$.
Then the Cayley graph $\Gamma = \Gamma(G, \Omega)$ has vertex-set G and, for g, h $\in$ G, $\{g, h\}$ is an edge of Γ if and only if $g^{-1}h \in \Omega$.

For example, the complete graph K_n is a Cayley graph $\Gamma(G, \Omega)$ for any group G of order n, where $\Omega = G - \{1\}$. As a less trivial example, let P_n denote the graph of the n-sided prism; then P_n is $\Gamma(G, \Omega)$, where G is the dihedral group D_{2n} and Ω consists of a pair of 'rotations' a, a^{-1}, and a 'reflection' b, satisfying $a^n = b^2 = (ab)^2 = 1$. The case $n = 5$ is illustrated in Fig. 15.

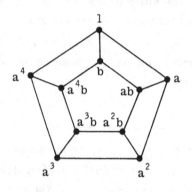

Fig. 15

We observe that a Cayley graph $\Gamma = \Gamma(G, \Omega)$ is always connected, by 5.3.1(i): let g, $h \in V$, and choose ω_1, ω_2, ..., $\omega_k \in \Omega$ such that $g^{-1}h = \omega_1 \omega_2 \ldots \omega_k$; then $(g = g_0, g_1, \ldots, g_k = h)$ is a walk in Γ, where $g_i = g_0 \omega_1 \ldots \omega_i$, $1 \leq i \leq k$. Condition (ii) of 5.3.1 ensures that no vertex is adjacent to itself, while (iii) guarantees that the edges are unordered pairs.

5.3.2 Theorem. A Cayley graph $\Gamma = \Gamma(G, \Omega)$ is vertex-transitive; in fact Aut Γ contains a regular subgroup isomorphic to G.

Proof. Each $g \in G$ induces the permutation $\tilde{g}$ of $V\Gamma = G$ given by $\tilde{g}(h) = gh$. Then $\tilde{g} \in$ Aut Γ, since if $\{h, k\} \in E\Gamma$, then $h^{-1}k \in \Omega$; thus $(gh)^{-1}gk \in \Omega$, and $\{gh, gk\} = \{\tilde{g}(h), \tilde{g}(k)\}$ is in $E\Gamma$. Clearly $\{\tilde{g} | g \in G\}$ is the subgroup required. //

The converse of this theorem is false; for example, Petersen's graph (Fig. 6) is vertex-transitive, but is not a Cayley graph for either group of order 10.

116

Now, let $M = (\Gamma, \rho)$ be a map, where Γ is a Cayley graph $\Gamma(G, \Omega)$. In this case, the vertex rotations ρ_v, $v \in V\Gamma = G$, can be regarded as permutations not only of the set $N(v)$, but also of the generating set Ω; thus one vertex can be more readily compared with another. We shall concentrate on the case when the induced permutations of Ω are all the same.

5.3.3 **Definition.** Let $r : \Omega \to \Omega$ be a cyclic permutation. Then the <u>Cayley map</u> $M(G, \Omega, r)$ is the map (Γ, ρ), where $\Gamma = \Gamma(G, \Omega)$ and, for $g \in G$ and $h \in N(g)$,

$$\rho_g(h) = g\,r(g^{-1}h).$$

Thus the group structure is used to determine the vertex rotations. In Fig. 16 below, where $r = (\omega_1 \omega_2 \ldots \omega_k)$, the edge $\{g, g\omega_1\}$ is determined by $\omega_1 \in \Omega$, so that the image of $g\omega_1$ under ρ_g is determined by $r(\omega_1) = \omega_2 \in \Omega$.

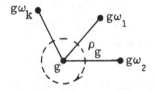

Fig. 16

For a specific example, consider the toroidal map for K_7 given in Section 5.1. This is a Cayley map $M(\mathbb{Z}_7, \mathbb{Z}_7^*, r)$, with $r = (1\,3\,2\,6\,4\,5)$. Then for $1 \le i \le 7 = 0$, $\rho_i(g) = i + r(g - i)$, where we use additive notation in the abelian group; that is, $\rho_i = (1 + i\ \ 3 + i \ldots 5 + i)$. On the other hand, the graph in Fig. 7(a) is a Cayley graph $\Gamma(\mathbb{Z}_2^3, \Omega)$ with Ω a set of three independent elements of order two, but the corresponding map of genus zero is not a Cayley map, since no suitable permutation r of Ω can be found.

In Theorem 5.3.2 we say that Cayley graphs are vertex-transitive. In fact, when $\Gamma = \Gamma(G, \Omega)$, the regular subgroup of Aut Γ isomorphic to G is also a subgroup of Aut M, for $M = M(G, \Omega, r)$.

5. 3. 4 **Theorem.** <u>A Cayley map</u> $M(G, \Omega, r)$ <u>is vertex-transitive.</u>
(<u>In fact</u>, Aut $M(G, \Omega, r)$ <u>contains a regular subgroup isomorphic to</u> G.)

Proof. For each $g \in G = V\Gamma$, define $\tilde{g} : V\Gamma \to V\Gamma$ by $\tilde{g}(h) = gh$.
(The isomorphism $g \mapsto \tilde{g}$ defines the left-regular representation of G
first noted by Cayley and used in 5. 3. 2 to show that $\Gamma(G, \Omega)$ is vertex-
transitive.) It remains only to show that $\tilde{g} \in$ Aut M; that is, that
$\rho_{\tilde{g}(h)} = \tilde{g}\rho_h\tilde{g}^{-1}$, for each $h \in V\Gamma$. Let $g' \in N(\tilde{g}(h))$; then

$$\rho_{\tilde{g}(h)}(g') = \rho_{gh}(g')$$
$$= gh \ r((gh)^{-1}g')$$
$$= \tilde{g}(h \ r(h^{-1}g^{-1}g'))$$
$$= \tilde{g}\rho_h(g^{-1}g')$$
$$= \tilde{g}\rho_h\tilde{g}^{-1}(g'). \ //$$

We remark that if Γ is a Cayley graph $\Gamma(G, \Omega)$ and if $M = (\Gamma, \rho)$
is <u>not</u> a Cayley map $M(G, \Omega, r)$, then M could be vertex-transitive (as
in Fig. 11 for $\Gamma = K_4$, with $G = \mathbb{Z}_4$), but that Aut M does not contain
a left-regular representation of G.

Cayley maps $M = M(G, \Omega, r)$ need not be symmetrical; in many
cases Aut M will consist cnly of the automorphisms $\tilde{g}$ just constructed,
so that Aut M $\approx$ G. For example, the toroidal map for K_4 given in
Fig. 14(a) has this property, where $G = \mathbb{Z}_4$, $\Omega = G^*$, and $r = (1 \ 2 \ 3)$.
In fact, this example generalizes as follows.

5. 3. 5 **Theorem.** <u>Let</u> $M(G, \Omega, r)$ <u>be a Cayley map which is not</u>
<u>symmetrical, and let</u> $|\Omega|$ <u>be prime; then</u> Aut M $\approx$ G.

Proof. By 5. 2. 6, $|A_v|$ divides $|\Omega|$, where A = Aut M and v is
any vertex of $\Gamma(G, \Omega)$. By 5. 3. 4, M is vertex-transitive, so that by
1. 3. 2 $|A| = |G| \cdot |A_v|$, where $|A_v| = 1$ or $|\Omega|$, since $|\Omega|$ is prime.
But $|A_v| \neq |\Omega|$, since M is not symmetrical; thus $|A| = |G|$, and
A = G. //

118

We now introduce a condition which is sufficient for the automorphism group of $M(G, \Omega, r)$ to be strictly larger than G.

5. 3. 6 Theorem. <u>Let $M = M(G, \Omega, r)$ be a Cayley map and let $\alpha \in$ Aut G be a group automorphism such that the restriction of α to Ω is a power r^l, where $1 \le l < |\Omega|$. Then α is a map automorphism fixing 1.</u>

Proof. Since α is a group automorphism, $\alpha(1) = 1$. Suppose $\{u, v\} \in E\Gamma$, so that $u^{-1}v \in \Omega$; thus $\alpha(u^{-1}v) = \alpha(u)^{-1}\alpha(v) \in \Omega$, $\{\alpha(u), \alpha(v)\} \in E\Gamma$, and α is an automorphism of the Cayley graph $\Gamma(G, \Omega)$. To show that α is in fact a map automorphism, we verify that $\rho_{\alpha(g)} = \alpha \rho_g \alpha^{-1}$, for all $g \in G$:

$$
\begin{aligned}
\rho_{\alpha(g)}(h) &= \alpha(g) \, r((\alpha(g))^{-1}h) \\
&= \alpha(g) \, r(\alpha(g^{-1})h) \\
&= \alpha(g) \, r^{l+1}(g^{-1}\alpha^{-1}(h)) \\
&= \alpha(g \, r(g^{-1}\alpha^{-1}(h)) \\
&= \alpha \rho_g \alpha^{-1}(h). \quad /\!/
\end{aligned}
$$

In the special case when $l = 1$ in 5. 3. 6, the automorphism group of $M(G, \Omega, r)$ is as large as is possible, as we now show.

5. 3. 7 Theorem. <u>Let $M = M(G, \Omega, r)$ be a Cayley map, with $\alpha \in$ Aut G such that the restriction $\alpha|\Omega = r$. Then M is a symmetrical map.</u>

Proof. By 5. 3. 6, $\alpha \in (\text{Aut } M)_1$. By 5.2.6, $|(\text{Aut } M)_1|$ divides $\kappa = |\Omega|$; but since $\alpha|\Omega = r$ and r is a κ-cycle, $|(\text{Aut } M)_1| = \kappa$ (in fact, $(\text{Aut } M)_1$ is generated by α). Next, by 5.3.4 Aut M is transitive on $V\Gamma$, where $\Gamma = \Gamma(G, \Omega)$, so that 1.3.2 gives

$$
\begin{aligned}
|\text{Aut } M| &= |V\Gamma| \cdot |(\text{Aut } M)_1| \\
&= |V\Gamma| \cdot \kappa = 2|E\Gamma|,
\end{aligned}
$$

and M is symmetrical. $/\!/$

For example, the toroidal map $M(\mathbf{Z}_5, \mathbf{Z}_5^*, r)$, $r = (1\ 2\ 4\ 3)$, is symmetrical, since $\alpha = (0)(1243)$ is an automorphism of $\mathbf{Z}_5$. (See Fig. 17.)

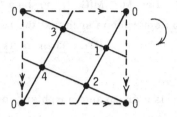

Fig. 17

We conclude this section by calculating the genus of an arbitrary Cayley map $M(G, \Omega, r)$. Recall from 5.1 that the number of faces of the map $M(\Gamma, \rho)$ is given by the number of orbits of the permutation ρ^*, acting on the sides of Γ by $\rho^*(u, v) = (v, \rho_v(u))$. This action induces a permutation (not necessarily cyclic) of Ω in the following way. A side $(v, v\omega)$, corresponding to the generator ω, is followed under ρ^* by the side $(v\omega, \rho_{v\omega}(v))$ (Fig. 18) and

$$\rho_{v\omega}(v) = v\omega\, r((v\omega)^{-1}v) = v\omega\, r(\omega^{-1}).$$

Thus the side $\rho^*(v, v\omega)$ corresponds to the generator $r(\omega^{-1})$. That is, the permutation of Ω induced by ρ^* is given by $\bar{r}(\omega) = r(\omega^{-1})$. Now let $\bar{r}$ have t cycles $\Omega_1, \Omega_2, \ldots, \Omega_t$ in its action on Ω, with m_i the order of $\omega_{i1}\omega_{i2} \ldots \omega_{ik_i}$ in G, where $\Omega_i = (\omega_{i1}\omega_{i2} \ldots \omega_{ik_i})$, $1 \le i \le t$. We note that m_i is well-defined, as a change in the initial member of the cycle Ω_i yields a conjugate element of G. Each m_i is said to be a period of the map M. The <u>length</u> of a face is the number of sides occurring in the corresponding cycle of ρ^*.

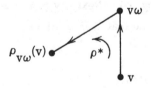

Fig. 18

5.3.8 Theorem. <u>Let</u> m_1, m_2, $\ldots$, m_t <u>be the periods of the</u> <u>Cayley map</u> $M(G, \Omega, r)$; <u>then the genus</u> g <u>of</u> M <u>is given by</u>

$$4(g - 1) = |G|(|\Omega| - 2 - 2\sum_{i=1}^{t} 1/m_i).$$

Proof. Consider $\omega_{i1} \in \Omega$, and let $\Omega_i = (\omega_{i1} \omega_{i2} \ldots \omega_{ik_i})$ be a cycle of $\bar{r}$. The cycle of ρ^* beginning with a side $(v, v\omega_{i1})$ of $\Gamma(G, \Omega)$ corresponding to ω_{i1} is only completed for the first non-trivial word θ taking the ω_{ij}'s in sequence and satisfying $(v\theta, v\theta\omega_{ij}) = (v, v\omega_{i1})$. Thus $\theta = 1 = (\omega_{i1} \omega_{i2} \ldots \omega_{ik_i})^{m_i}$. This corresponds to a face of length $m_i k_i$. Moreover, if we let ϕ_i denote the number of such faces f and use 1.2.2 to count the set $\{(v, f) | v \in f\}$ in two ways, observing that each vertex v of G occurs precisely k_i times in this process, then we find that $\phi_i(m_i k_i) = |G|k_i$. Thus, for $1 \leq i \leq t$, the cycle Ω_i corresponds to $|G|/m_i$ faces, each of length $m_i k_i$. The result now follows from the formula 5.1.6,

$$|V| - |E| + |F| = 2 - 2g,$$

using $|V| = |G|$, $2|E| = |G||\Omega|$, and $|F| = \sum_{i=1}^{t} |G|/m_i$. $/\!/$

In the special case where each $m_i = 1$, we have

5.3.9 $4(g - 1) = |G|(|\Omega| - 2(1 + t))$. $/\!/$

For example, in the map $M(\mathbb{Z}_5, \mathbb{Z}_5^*, r)$, $r = (1243)$, given in Fig. 17, $\bar{r} = (1342)$; thus $t = 1$, $m_1 = 1$, and we check that $g = 1$. In the map $M(\mathbb{Z}_7, \mathbb{Z}_7^*, r)$, $r = (132645)$, of Section 5.1, $\bar{r} = (142)(356)$, so that $t = 2$, $m_1 = m_2 = 1$, and again $g = 1$.

On the other hand, if (for example) $M = M(A_4, \Omega, r)$, with $\Omega = \{(123), (321), (12)(34)\}$ and $r = ((123), (321), (12)(34))$, then $\bar{r} = ((123), (12)(34))((321))$; thus $t = 2$, $m_1 = m_2 = 3$, and we use 5.3.8 to find $g = 0$ (see Fig. 19). Similarly, the 'soccer ball' map $M(A_5, \Omega, r)$, $\Omega = \{(12345), (54321), (12)(34)\}$, has $t = 2$, $m_1 = 3$, $m_2 = 5$ and - of course - $g = 0$.

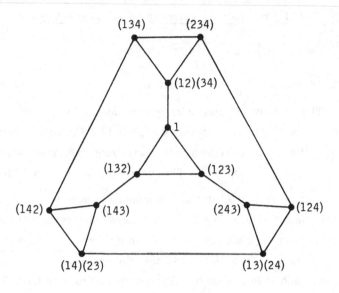

Fig. 19

5.4 Complete maps and a theorem of Frobenius

A <u>complete map</u> is a map whose underlying graph is a complete
graph K_n. In intuitive terms, it corresponds to a system of points on
a surface, each of which is joined to all the others in such a way that
there are no 'crossings'. One of the first aspects of this situation to be
studied was a version of the Heawood map-colouring problem: Given n,
what is the smallest value of g for which there is a rotation ρ on K_n
such that the genus of the map (K_n, ρ) is g? Our interests lie in a
different direction. We shall be concerned with maximizing the symmetry
of the map, rather than minimizing its genus, and this will lead us to
some classical results of permutation-group theory.

We begin by drawing attention to three examples of complete maps.
First, we have the map (K_4, ρ) of Fig. 11. Secondly, there is a map
(K_5, ρ) of genus 1, which is represented in the conventional way as a
drawing on a torus in Fig. 17. Finally, the rotation on K_7 tabulated in
Section 5.1 also gives rise to a map with genus 1, and this is drawn in
Fig. 20.

122

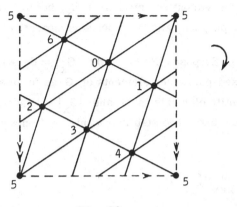

Fig. 20

These three are symmetrical maps, and they are the simplest examples
of a general construction for symmetrical complete maps.

Our first remark is simple, but fundamental.

5.4.1 Lemma. Suppose that ρ is a rotation on the complete graph
K_n and G is the automorphism group of the map (K_n, ρ). Then no non-
identity element of G can fix more than one vertex.

Proof. Any two distinct vertices of K_n are adjacent, and so the
result is an immediate consequence of Lemma 5.2.5. //

The preceding observation leads directly to a class of permutation
groups which have been the subject of much study. The transitive permutation group (G, X) is said to be a Frobenius group if the identity is
the only element of G which has more than one fixed point. Thus,
Lemma 5.4.1 says that the automorphism group of a vertex-transitive
complete map is a Frobenius group. The most important fact about a
Frobenius group is that it possesses a regular normal subgroup.

5.4.2 Theorem. Let (G, X) be a Frobenius group, and suppose
that N^* is the set of fixed-point-free elements of G, and $N = N^* \cup \{1\}$.
Then

    (i) $|N| = |X|$;

    (ii)   if the stabilizer G_x is abelian, N is a regular normal subgroup of (G, X).

[In fact, (ii) is true without restriction on G_x, but the special case is sufficient for our purposes and its proof is comparatively elementary.]

Proof. (i) Suppose $|X| = n$ and $|G_x| = d$. We may calculate the number of fixed-point-free elements of G as follows. There are $|G| - 1$ non-identity elements in G and $|G_x| - 1$ non-identity elements in each stabilizer; also, two stabilizers intersect only in the identity. Hence

$$N^* = G^* - \bigcup_{x \in X} G_x^* ,$$

$$|N^*| = (nd - 1) - n(d - 1) = n - 1.$$

Thus $|N| = |X|$, as required.

(ii)   We shall prove that N is a normal subgroup of G by constructing a homomorphism $t : G \to G_x$ with kernel N. Since G acts transitively, we may choose a family $\{l_y\} \subseteq G$ such that $l_y(x) = y$ ($y \in X$). For each g in G and y in X, define

$$g_y = l_{gy}^{-1} g\, l_y.$$

Then $g_y(x) = x$, that is, $g_y \in G_x$. Since G_x is abelian, the order of composition of its members is unimportant, and the following product is meaningful:

$$t(g) = \prod_{y \in X} g_y.$$

To show that $t : G \to G_x$ is a homomorphism, we argue as follows, remembering that G_x is abelian:

$$t(gh) = \prod_{y \in X} (gh)_y$$

$$= \prod_{y \in X} l_{ghy}^{-1} \cdot gh \cdot l_y$$

$$= \prod_{y \in X} (l_{ghy}^{-1} \cdot g \cdot l_{hy}) \cdot (l_{hy}^{-1} \cdot h \cdot l_y)$$

$$= \prod_{y \in X} (l_{ghy}^{-1} \cdot g \cdot l_{hy}) \cdot \prod_{y \in X} (l_{hy}^{-1} \cdot h \cdot l_y).$$

Putting $z = hy$ in the first product, we have

$$t(gh) = \prod_{z \in X} g_z \cdot \prod_{y \in X} h_y = t(g).t(h).$$

Now we need a couple of remarks about the cycle structure of elements of G. If $g \in N$ and a shortest cycle of g has length $\theta > 1$, then g^θ fixes more than one point and so $g^\theta = 1$. Thus all the cycles of a fixed-point-free permutation in G have equal length. Similarly, if $g \in G_x$, then g has one cycle (x) of length 1 and all the remaining cycles have equal length.

Suppose that $(y\,z\,\ldots\,v\,w)$ is a θ-cycle of g. Then

$$g_w g_v \cdots g_z g_y = l_y^{-1} g\, l_w \cdot l_w^{-1} g\, l_v \cdots l_z^{-1} g\, l_y$$

$$= l_y^{-1} g^\theta l_y .$$

So, when $g \in N$, we have $g^\theta = 1$, and consequently $g_w g_v \cdots g_z g_y = 1$. It follows that for $g \in N$ we have $t(g) = 1$, and so $N \subseteq \mathrm{Ker}\ t$. Further, if $g \notin N$, then $g \in G_y$ say, and g permutes $X - \{y\}$ in cycles of equal length ψ, so that $g^\psi = 1$. In this case we have

$$t(g) = g_y = l_{gy}^{-1} g\, l_y = l_y^{-1} g\, l_y \neq 1.$$

This completes the proof that N is the kernel of the homomorphism t, and so $N \trianglelefteq G$. Since $|N| = |X|$ and N* consists of fixed-point-free permutations, N must act regularly on X. //

The preceding result enables us to classify, in a sense, all vertex-transitive complete maps. First, we recall the Cayley map construction introduced in the previous section. If the generating set Ω is the whole of $G^* = G - \{1\}$, then the Cayley graph $\Gamma(G, \Omega)$ is necessarily a complete graph K_n with $n = |G|$. Choosing any cyclic permutation r of the $n - 1$ members of G* gives rise to a Cayley map $M(G, G^*, r)$; this map is complete, and vertex-transitive (by 5.3.4). The usefulness of the Frobenius theorem lies in the fact that it provides a converse to this simple observation.

5.4.3 Theorem. <u>Suppose that the automorphism group of a complete map</u> M <u>acts transitively on the vertices. Then</u> M <u>is a Cayley map.</u>

Proof. It follows from Lemma 5.4.1 that $A = \text{Aut } M$ is a Frobenius group. Since the stabilizer of a vertex is necessarily cyclic (5.2.6), the second part of Theorem 5.4.2 tells us that A has a regular normal subgroup N. The vertices of M are in one-to-one correspondence $y \longleftrightarrow n_y$ with the elements of N, where n_y is the unique member of N for which $n_y(x) = y$. (x is an arbitrary, but fixed, vertex of M.)

Let ρ_x be the rotation at x in M. Since M is complete, ρ_x is a cyclic permutation of the remaining vertices. The correspondence $y \longleftrightarrow n_y$ induces a cyclic permutation of N^*, defined by

$$(*) \qquad r(n_a) = n_b \Longleftrightarrow \rho_x(a) = b.$$

Thus we have a Cayley map $\overline{M} = M(N, N^*, r)$. In order to show that $\overline{M}$ is isomorphic to the given map M, we have to check that the given rotation ρ in M induces the rotation $\overline{\rho}$ in $\overline{M}$ under the correspondence $y \longleftrightarrow n_y$. In other words, we must show that

$$(**) \qquad \rho_y(z) = w \Rightarrow \overline{\rho}_{n_y}(n_z) = n_w.$$

Let $n_y^{-1} n_z = n_p$, and suppose $\rho_x(p) = q$ so that, by $(*)$, $r(n_p) = n_q$. The definition of $\overline{\rho}$ (5.3.3) gives

$$\overline{\rho}_{n_y}(n_z) = n_y \, r(n_y^{-1} n_z) = n_y \, r(n_p) = n_y n_q \; .$$

Now,

$$
\begin{aligned}
n_y n_q(x) &= n_y(q) = n_y \rho_x(p) = n_y \rho_x n_p(x) \\
&= n_y \rho_x n_y^{-1} n_z(x) \\
&= n_y \rho_x n_y^{-1}(z).
\end{aligned}
$$

Since n_y is an automorphism of M, we have

$$n_y \rho_x n_y^{-1} = \rho_{n_y}(x) = \rho_y.$$

Thus $n_y n_q(x) = \rho_y(z) = w$, and since N acts regularly, $n_y n_q = n_w$.
This completes the proof of (**). //

The examples involving K_4, K_5, and K_7 listed at the beginning
of this section are all vertex-transitive, and hence they are Cayley maps.
The relevant groups, which occur as regular normal subgroups of the
automorphism groups, are $\mathbb{Z}_2^2$, $\mathbb{Z}_5$, and $\mathbb{Z}_7$ respectively. As we have
already remarked, these examples are in fact symmetrical maps.

Let us summarise the facts about the automorphism group A of
a symmetrical complete map which follow from the general theory.

(i) $|A| = n(n - 1)$ (from Definition 5.2.7);

(ii) A acts sharply 2-transitively on the vertices (since each
ordered pair of distinct vertices is a side, and A acts regularly on the
set of sides);

(iii) A is a Frobenius group;

(iv)   the stabilizer of a vertex is a cyclic group of order $n - 1$.
It is a consequence of (iii), (iv), and Theorem 5.4.2 that A has a
regular normal subgroup. More generally, we shall show that the same
conclusion follows from (ii) alone. This result is not only an important
theorem in the general theory of permutation groups, it is also very
significant in the theory of symmetrical complete maps, since the proof
yields more information than we have so far established.

5.4.4 Theorem. <u>If</u> (G, X) <u>is a sharply</u> 2-<u>transitive group, then</u>
$|X|$ <u>is a prime power,</u> p^r. <u>The fixed-point-free elements of</u> G, <u>to-</u>
<u>gether with the identity, form a regular normal subgroup which is also</u>
<u>a Sylow</u> p-<u>subgroup of</u> G.

Proof. Let $|X| = n$, and suppose N^* is the set of fixed-point-free
elements of G. From Theorem 5.4.2, we know that $|N^*| = n - 1$. Let
p be a prime dividing n and g an element of order p in G. A cycle
of g has length 1 or p, and there is at most one cycle of length 1.
Since p divides n (the sum of the cycle lengths), g can have no cycles
of length 1. That is, $g \in N^*$.

Let K denote the conjugacy class of g in G. We have

$$|K| = |G : C(g)| = n(n - 1)/|C(g)| .$$

Now if $h \in C(g)$ and h fixes x, then h also fixes $g(x) \neq x$; thus $h = 1$. Hence $C(g)$ is a subset of $N^* \cup \{1\}$ and so $|C(g)| \leq n$. The equation for $|K|$ yields $|K| \geq n - 1$. But $K \subseteq N^*$ (since $g \in N^*$) and $|N^*| = n - 1$, so that $K = N^*$.

Now K was defined in terms of p, and N^* was not; thus p must be the only prime divisor of n, and $n = p^r$, say.

Let P be a Sylow p-subgroup of G. Since p does not divide $n - 1$, $|P| = p^r$. If $h \in P^*$, then considering the cycle lengths of h (as for g above), we see that h is fixed-point-free. Thus $P^* \subseteq N^*$, and since $|P^*| = |N^*| = n - 1$, we must have $P^* = N^*$. This shows that $N = N^* \cup \{1\}$ is a group: in fact, $N = P$. N is normal since it is the union of two conjugacy classes, $\{1\}$ and K. //

The theorem enables us to amplify the results about sharply 2-transitive groups obtained in Section 2.8. We remark first that (by 1.7.6(i)), the regular normal subgroup N must be an elementary abelian group. This suggests the possibility of a further relationship, beyond that established in Section 2.8, between near-fields and sharply 2-transitive groups. We already know (2.8.1) that the affine transformations of a near-field form a sharply 2-transitive group; Theorem 5.4.4 tells us that the group has degree p^r and so the near-field must have p^r elements. Furthermore, if a sharply 2-transitive group (G, X) is given, we have an additive structure on X induced by the action of the regular elementary abelian subgroup N. In fact, it is possible to define a multiplication on X so that it becomes a near-field. Thus, near-fields and sharply 2-transitive groups are essentially the same thing.

We now return to the study of complete maps. Theorem 5.4.4 implies that a symmetrical map (K_n, ρ) can occur only when n is a prime power. Examples for the cases $n = 4, 5, 7$ have already been given, and we now show that these examples are typical.

5.4.5 Theorem. <u>There is a rotation ρ on K_n which gives rise to a symmetrical map if and only if n is a prime power.</u>

Proof. Since the automorphism group of a symmetrical complete map is sharply 2-transitive, n must be a prime power, by 5.4.4.

For the converse, we have to construct a symmetrical map for K_n when n is a prime power. By 5.4.3, this must be done by using a Cayley map construction $M(G, G^*, r)$. Suppose $n = p^r$; take $G = (\mathbf{Z}_p)^r$, regarded as the underlying additive group of the field $GF(p^r)$. The generating set must be G^*, and the permutation $r : G^* \to G^*$ is given by

$$r(x) = tx \qquad (x \in G^*),$$

where t is a primitive element of $GF(p^r)$, and the multiplication is the usual field multiplication. The definition of a primitive element ensures that r is a cyclic permutation of G^*. Also, the distributive law implies that $r(x + y) = r(x) + r(y)$ so that r (extended by putting $r(0) = 0$) is an automorphism of the group G. It follows from Theorem 5.3.7 that $M(G, G^*, r)$ is a symmetrical map. //

It is instructive to examine the complete maps constructed above in more detail. We shall use the permutation $\bar{r}$ introduced in Section 5.3, remembering that the group G is written additively in this case, so that $\bar{r}(x) = -tx$. There are three possibilities, according as n is congruent to 0, 1, or 3 modulo 4.

$\underline{n = 2^m}$ In this case $\bar{r} = r$ and there is a single cycle

$$\Omega_1 = (1, t, t^2, \ldots, t^{n-2}).$$

Since $1 + t + t^2 + \ldots + t^{n-2} = 0$, the proof of Theorem 5.3.8 shows that the map has n faces, each with $n - 1$ sides. The genus is equal to $(n - 1)(n - 4)/4$.

$\underline{n \equiv 1 \pmod 4}$ In this case $(-t)^{(n-1)/2}$ is not equal to 1, and $\bar{r}$ once again has a single cycle

$$\Omega_1 = (1, -t, t^2, \ldots, (-t)^{n-2}).$$

Put $z = 1 - t + t^2 - \ldots + (-t)^{n-2}$. Then $z = -tz$ and $z = 0$ again. Thus we have a map with n faces, each having $n - 1$ sides, and the genus is $(n - 1)(n - 4)/4$.

$\underline{n \equiv 3 \pmod 4}$ Here we have $(-t)^{(n-1)/2} = 1$, so that $\bar{r}$ has two cycles:

$$\Omega_1 = (1, -t, \ldots, (-t)^{(n-3)/2}), \quad \Omega_2 = (-1, t, \ldots, -(-t)^{(n-3)/2}).$$

We find that $z_1 = -z_2 = tz_1$, so that $z_1 = z_2 = 0$. Thus the map has 2n faces, each with $\frac{1}{2}(n - 1)$ sides, and the genus is $(n^2 - 7n + 4)/4$.

Finally, we observe that these maps give rise to 2-designs, $(X, \mathfrak{B})$. We take X to be the set of vertices and $\mathfrak{B}$ to be the set of faces, a face being regarded as a set of vertices (there are no repetitions here). In the first two cases treated above the associated design is a complete design; in the third case we obtain a $2 - (n, \frac{1}{2}(n - 1), 2)$ design which decomposes naturally into two disjoint $2 - (n, \frac{1}{2}(n - 1), 1)$ symmetric designs. When $n = 7$ (Fig. 20), we obtain the configuration which has recurred throughout these notes, as a projective plane, a Steiner triple system, a Hadamard design, and now, as a complete map.

5.5 Symmetrical maps

Among the many questions which arise in connection with symmetrical maps, there are two very obvious ones:

(i)   Given a graph Γ, what are the symmetrical maps (Γ, ρ)?

(ii)   Given a non-negative integer g, what are the possible symmetrical maps of genus g?

In Section 5.4 we discussed the first question in the case $\Gamma = K_n$; some other graphs and their symmetrical maps will be treated in Sections 5.6, 5.7, and 5.8.

The second question has the longer pedigree: in the case $g = 0$, it was solved by the ancient Greeks, since the symmetrical maps of genus zero are just the regular polygons and the classical regular polyhedra. We shall discuss a few basic aspects of the problem, using simple numerical conditions.

We have already remarked, at the end of Section 5.2, that there are two constants κ and l associated with a symmetrical map. The number κ is the valency of the graph, and l is the number of sides in each face of the map.

5. 5. 1 Theorem. Let M be a symmetrical map, with V, E, F
denoting its vertices, edges, and faces. Suppose κ and l are as
defined above, and $g = g(M)$. Then

(i) $\kappa|V| = l|F| = 2|E|$;

(ii) $g = 1 + (\dfrac{(\kappa - 2)(l - 2) - 4}{4l})|V|$.

Proof. The equations in part (i) are a direct consequence of the
basic counting principle (1. 2. 2).

For part (ii), we substitute for $|E|$ and $|F|$ in the formula for
the genus: $|V| - |E| + |F| = 2 - 2g$. //

When $\kappa = 2$, we must have $l = |V|$ and $g = 0$; the map corres-
ponds to a single polygon drawn on a sphere. In the following discussion
we shall assume $\kappa \geq 3$ (and $l \geq 3$).

The two parts of Theorem 5. 5. 1 imply that, provided $g \neq 1$, the
numbers κ, l and g determine $|V|$, $|E|$, and $|F|$. For example,
when $g = 0$ we must have

$$(\kappa - 2)(l - 2) < 4,$$

leading to the five possibilities $(\kappa, l) = (3, 3), (3, 4), (3, 5), (4, 3),$
$(5, 3)$. In each case $|V|$, $|E|$, and $|F|$ are determined, and there is
a unique map, corresponding to one of the five regular convex polyhedra.

The case $g = 1$ is exceptional. We must have

$$(\kappa - 2)(l - 2) = 4,$$

leading to the three possibilities $(\kappa, l) = (3, 6), (4, 4), (6, 3)$. Here
the numbers $|V|$, $|E|$, and $|F|$ are not determined, and there is, in
fact, an infinite family of symmetrical maps in each of the three cases.
The maps may be obtained by taking one of the three regular tessellations
of the plane (hexagonal, square, or triangular) and making identifications
on a suitable portion of it so that a torus is formed. There is an infinite
number of ways of doing this in each case.

When $g \geq 2$, we return to the situation where $|V|$, $|E|$, $|F|$ are
determined by κ, l, and g. Furthermore, for each $g \geq 2$, the number

of symmetrical maps of genus g is finite.

5. 5. 2 Theorem.   Let M be a symmetrical map of genus $g \geq 2$.
Then

$$|\text{Aut } M| \leq 84(g - 1),$$

and equality holds if and only if $(\kappa, l) = (7, 3)$ or $(3, 7)$.

Proof.   By definition, M is symmetrical if and only if
$|\text{Aut } M| = 2|E|$. Using both parts of Theorem 5. 5. 1, we obtain the
following equation linking $2|E|$ and g:

$$2|E| = (\frac{4\kappa l}{(\kappa - 2)(l - 2) - 4})(g - 1).$$

Since we are assuming that $\kappa \geq 3$ and $l \geq 3$, the coefficient of g - 1
has a maximum value of 84 and this occurs when $(\kappa, l) = (7, 3)$ or
(3, 7). //

The simplest example of a symmetrical map with 84(g - 1) automorphisms is the so-called Klein map of genus 3, whose automorphism
group is the simple group PSL(2, 7). In general, it is not known for
which values of g a symmetrical map with 84(g - 1) automorphisms
exists: there is none for g = 2, 4, 5, 6; but there is one for g = 7.
This is just one of many open questions concerning symmetrical maps;
it is an area which seems to offer great possibilities for the interplay of
combinatorial, group-theoretical, and geometrical ideas.

---

5. 6   Project: Generalized Cayley maps

Let $\Gamma = \Gamma(G, \Omega)$ be a Cayley graph and suppose G has a subgroup
H of index n. Let $g_1 = 1$, g_2, ..., g_n be a family of right coset representatives for H in G, and let r_1, r_2, ..., r_n be cyclic permutations
of Ω. Define a rotation ρ on Γ as follows:

$$\rho_g(k) = g \, r_i(g^{-1}k) \quad (g \in Hg_i, \, \{g, k\} \in E\Gamma).$$

The map $M = (\Gamma, \rho)$ is a <u>Cayley map of index</u> n. The theory of such maps can be developed along similar lines to that for ordinary Cayley maps (which are simply the case $n = 1$).

5.6.1 Show that the automorphism group of M contains a subgroup isomorphic to H.

5.6.2 Show that, if α is an automorphism of G fixing each right coset of H in G setwise and whose restriction to Ω is r_1, and if r_1 commutes with each r_i, then α is an automorphism of M, fixing the identity vertex. Thus $|\text{Aut } M| \geq |H| |\Omega|$.

5.6.3 Suppose that $\Gamma = \Gamma(G, \Omega)$ is a Cayley graph and that G contains a subgroup H of index 2 such that $H \cap \Omega = \emptyset$. Prove that Γ is a <u>bipartite</u> graph: that is, the vertices of Γ can be partitioned into two disjoint sets V_1 and V_2 in such a way that every edge has one vertex in V_1 and one vertex in V_2.

Let M be the Cayley map of index 2 obtained from Γ by using the subgroup H and permutations r_1, r_2 satisfying $r_2(\omega)^{-1} = r_1^{-1}(\omega^{-1})$. Find the analogue of the formula 5.3.8 for M.

5.6.4 Let G be the group $\mathbf{Z}_2^n$, whose members are the n-tuples $(x_1, x_2, \ldots, x_n)$, $x_i = 0$ or 1, and with the group operation taken to be addition. Let $\Omega = \{e_i | 1 \leq i \leq n\}$, where e_i has 1 in the ith coordinate and zeros elsewhere. The Cayley graph $\Gamma(G, \Omega)$ is the n-<u>cube</u> Q_n; the case $n = 3$ is illustrated in Fig. 7(a).

Find a subgroup H of G satisfying the conditions of 5.6.3, and let r_1 and r_2 be as stated there, with $r_1(e_i) = e_{i+1}$ (subscripts mod n). Show that the genus of the resulting Cayley map of index 2 is

$$g = 1 + 2^{n-3}(n - 4) \qquad (n \geq 2).$$

5.6.5 Show that the genus of any map (Q_n, ρ) cannot be less than $1 + 2^{n-3}(n - 4)$: this number is the <u>genus</u> of the <u>graph</u> Q_n, denoted by $\gamma(Q_n)$.

5.6.6   Let $M = (Q_n, \rho)$ as in 5.6.4. Show that $\beta : VQ_n \to VQ_n$, given by $\beta(a_1, a_2, \ldots, a_n) = (a_n, \ldots, a_2, a_1 + 1)$, is in Aut M, and deduce that M is vertex-transitive. Use 5.6.2 to conclude that M is symmetrical.

5.6.7   Let G be the additive group of integers modulo 2n, $\Omega = \{1, 3, \ldots, 2n - 1\}$ and $H = \{0, 2, \ldots, 2n - 2\}$. Check that $\Gamma(G, \Omega)$ is the complete bipartite graph $K_{n,n}$ as defined in 4.6.3. Let M be the Cayley map of index 2 constructed from Γ, H, and the permutations r_1, r_2 defined by $r_1(2j - 1) = r_2(2j - 1) = 2j + 1$. Show that M is a symmetrical map. (Note that M is in fact a Cayley map.)

5.7   Project: Paley maps

Let G be the group Z_p^r, regarded as the additive group of the field $GF(q)$, $q = p^r$, and let $\Omega \subseteq G$ be the set of non-zero squares x^2, $x \in GF(q)$. If $q \equiv 1 \pmod 4$, then Ω satisfies the three conditions stated in 5.3.1 (recall that G is written additively here). In this case, the Paley graph $P(q)$ is defined to be the Cayley graph $\Gamma(G, \Omega)$, and the Paley map $MP(q)$ is defined to be the Cayley map $M(G, \Omega, r)$, where $r(x) = t^2 x$ for some fixed primitive element t of $GF(q)$. In Fig. 21 we show the Paley map $MP(9)$.

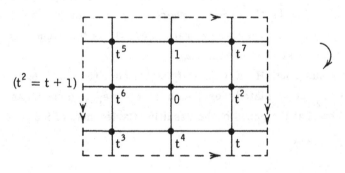

Fig. 21

5.7.1   Show that a Paley graph admits a rank 3 group of automorphisms, acting transitively on the vertices, and find its parameters as a strongly regular graph.

5.7.2   Show that $P(q)$ is isomorphic to its complement $P(q)^c$.

5.7.3   Prove that the Paley maps are symmetrical maps, and that Aut $MP(q)$ is a subgroup of the affine group of transformations of $GF(q)$. Deduce that Aut $MP(q)$ is a Frobenius group.

5.7.4   Find the genus of $MP(q)$, distinguishing the two cases $q \equiv 1, 5 \pmod 8$.

5.7.5   Let $\Gamma = (V, E)$ be a graph which has an 'anti-automorphism' β; that is, β is a permutation of V such that $\{v, w\} \in E$ implies $\{\beta v, \beta w\} \notin E$. Suppose also that $M = (\Gamma, \rho)$ is a symmetrical map such that

(i) $\beta \alpha \beta^{-1} \in$ Aut M, for each $\alpha \in$ Aut M;

(ii) $\beta^2 \in$ Aut M.

Show that $|V|$ must be a prime power congruent to 1 modulo 4. Explain how the Paley maps provide a converse to this result.

5.7.6*   Show that $|\text{Aut } P(q)| = \tfrac{1}{2} rq(q - 1)$, where $q = p^r$.

5.8   Project: Symmetrical Cayley maps

In Section 5.4 we constructed symmetrical Cayley maps for complete graphs of prime power order, and in Section 5.7 the construction was modified slightly for the Paley graphs. In this project we construct additional classes of symmetrical Cayley maps. The first of these will be a second class of symmetrical maps for the n-cubes Q_n (compare 5.6.6); then we construct two classes $M(G, \Omega, r)$ for $G = S_n$ and one class for $G = A_n$ (n odd).

5.8.1   Let $G = \mathbb{Z}_2^n$, with Ω as in 5.6.4, and let $\alpha : G \to G$ be given by $\alpha(a_1, a_2, \ldots, a_{n-1}, a_n) = (a_n, a_1, a_2, \ldots, a_{n-1})$. Show that α is a group automorphism, that $r = \alpha|\Omega$ is a cyclic permutation, and conclude that $M(Q_n, \Omega, r)$ is symmetrical. Show that the genus of M is

$$1 + 2^{n-2}(n - 3) = \gamma(Q_{n+1}), \quad n \geq 1.$$

5.8.2   Recall from Section 1.5 that $\Omega_0 = \{(12), (23), \dots, (n-1\ n)\}$ generates $G = S_n$. Let $\Omega = \Omega_0 \cup \{(n\ 1)\}$, and consider the inner automorphism α of S_n corresponding to conjugation by $(1\ 2\ 3\ \dots\ n)$. Show that $r = \alpha|\Omega$ is a cyclic permutation of Ω, and conclude that $\Gamma(S_n, \Omega)$ has a symmetrical map, for all $n \geq 2$. What is the genus of $M(S_n, \Omega, r)$? Check that the case $n = 3$ gives a toroidal map for the complete bipartite graph $K_{3,3}$ (as in 5.6.7); illustrate this map with a diagram.

5.8.3   Show that $\Omega' = \{(12), (13), \dots, (1\ n)\}$ is also a generating set for S_n. Now find a symmetrical map for $\Gamma(S_n, \Omega')$, for all $n \geq 2$. What is the genus of the map so constructed?

5.8.4   Use 1.5.2 to show that $G = A_n$ is generated by the 3-cycles $(1\ 2\ k)$, $3 \leq k \leq n$. Now let n be <u>odd</u> and suppose Ω consists of the 3-cycles $(1\ 2\ k)$ and $(k\ 2\ 1)$ for $3 \leq k \leq n$, so that $\Gamma(G, \Omega)$ is a Cayley graph. Let α be the outer automorphism of A_n corresponding to conjugation by $(12)(34 \dots n)$. Show that $\alpha|\Omega = r$ is a cyclic permutation of Ω and thus $M(A_n, \Omega, r)$ is symmetrical.

NOTES AND REFERENCES FOR CHAPTER 5

A good introduction to the topology of surfaces is given in the little book by Fréchet and Fan [4]. The relationship with groups and graphs is discussed by White [9]. Complete maps and symmetrical maps are treated in [1].

There is a certain amount of 'folk-lore' concerning symmetrical maps. Many interesting examples are to be found in Coxeter and Moser's book [3]. Brahana [2] constructed, for all $n \geq 4$, symmetrical maps whose full automorphism groups are S_n and A_n respectively. An entirely different approach enabled Macbeath [7] to construct a map with $84(g - 1)$ automorphisms for infinitely many values of g. The many aspects of the theory of maps - involving ideas from complex analysis, group theory, and combinatorics - are tied together in a paper by Jones and Singerman [6].

The theory of 'current graphs' is a useful tool in the construction of maps: Ringel's book [8] uses this theory to solve the problem of finding the genus of K_n. The 'dual' theory of 'voltage graphs' [5] allows a topological interpretation in terms of branched covering spaces.

1.   Biggs, N. L. Cayley maps and symmetrical maps. Proc. Cambridge Phil. Soc., 72 (1972), 381-6.
2.   Brahana, H. R. Regular maps and their groups. Amer. J. Math., 49 (1927), 268-84.
3.   Coxeter, H. S. M. and Moser, W. O. J. Generators and Relations for Discrete Groups. (Springer-Verlag, 1965.)
4.   Fréchet, M. and Fan, K. Initiation to Combinatorial Topology. (Prindle, Weber and Schmidt, 1967.)
5.   Gross, J. L. Voltage graphs. Discrete Math., 9 (1974), 239-46.
6.   Jones, G. A. and Singerman, D. Theory of maps on orientable surfaces. Proc. Lond. Math. Soc., (3) 37 (1978), 273-307.
7.   Macbeath, A. M. On a theorem of Hurwitz. Proc. Glasgow Math. Assoc., 5 (1961), 90-6.
8.   Ringel, G. Map Color Theorem. (Springer-Verlag, 1974.)
9.   White, A. T. Graphs, Groups and Surfaces. (North-Holland, 1973.)

Index

action 1
adjacency matrix 90
adjacent 83
admissible 13
affine transformation 27
alternating group 2
automorphism (linear) 28
automorphism (semilinear) 44
automorphism of a design 65
automorphism of a graph 85
automorphism of a map 110
axis 45

bilinear form 45
bipartite graph 133
biplane 101
block 55
Burnside's Lemma 4

Cayley graph 115
Cayley map 117
centre 45
circuit 86
class equation 10
collineation 44
complement graph 85
complete bipartite geaph 99
complete design 56
complete graph 83

complete map 122
conjugate 2
connected 81, 83
contraction 67
coordinates 28
correlation 46
cube graph 133

degree 1
design 55
diagonal 13
digraph 81
dimension (projective) 37
distance-regular graph 101
divisibility conditions 57
double coset 9
dual design 61

edge 83
elation 45
equation of hyperplane 30
equivalent permutation groups 17
equivalent rotations 110
extension class 77
extension of a design 67
extension (one-point) 11

face 106
faithful 1
finite field 26

strongly regular graph 88

subspace 37

surface 103

Sylow p-subgroup 20

Sylow's Theorems 10, 20

symmetrical map 114

symmetric design 59

symmetric group 1

symmetric orbit 82

symplectic form 46

symplectic group 47

transitive 5

transposed orbit 82

transposition 2

transvection 31

unitary form 46

unitary group 47

valency 86

vertex 83

vertex-transitive graph 85

vertex-transitive map 113

walk 81, 83